Teaching Science in the '90s

Teaching Science in a Climate of Controversy first appeared in 1986. Since then, many changes have taken place throughout the world. In the United States, public schools are still embroiled in controversy, and teaching has not gotten any easier.

Budget constraints are forcing scientific and educational institutions to "re-evaluate." Indeed, "values" (once considered of importance only in philosophy and religion) seem to be on everybody's mind. Leaders of science rush to define ethical misconduct in science before some Congressional committee does it for them. School administrators seek some kind of common moral ground to guide them in diverse multicultural settings. The public, aware of such tensions, wonders *whose* values are being taught to their children.

Consider the range of reactions to earlier versions of this publication. Of the hundreds of teachers, students, and parents who returned the "Grade Us" postcards stapled in the back, three-quarters graded the book's contents as A+, A, or B. Some 12 percent rated it a total failure, about half of them evidently seeing it as "typical creationist propaganda," the other half as "typical evolutionist propaganda." The authors made a few changes in a second printing (1987) and rewrote parts of the text for greater clarity in a third printing (1989). Those changes took care of a few substantive criticisms, but nothing could be done to satisfy the nonscientific biases of negative critics.

The religious biases of many who attack the teaching of evolution have been thoroughly documented by historian Ronald Numbers in *The Creationists* (Knopf, 1992). The public is less inclined to suspect ideological bias in those who *defend* the teaching of evolution, but in many cases such biases are real, as argued by legal scholar Phillip Johnson in *Darwin on Trial* (IVP, 1991). Our emphasis is on teaching good science without entanglement in political, philosophical, or religious controversy, but we highly recommend both books. We would have added them to the "additional resources" in the Appendix, but we decided to reprint the whole text just as we wrote it in 1989. Despite ongoing scientific research on each of the four "open questions" we posed, our conclusions did not need to be changed in 1993.

Two features have been added, however: 1) **A Voice for Evolution** *As Science*, a 1991 resolution passed by the American Scientific Affiliation (back of this page); and 2) **Exercises to Teach Critical Thinking**, based on evolutionary evidences and inferences (an Addendum at the end of the book). We have also changed from a magazine format to a book format, which we hope will make *Teaching Science in a Climate of Controversy* more convenient to use and more widely available in bookstores.

—-*Committee for Integrity in Science Education*
American Scientific Affiliation
1993

A Spanish language version of *Teaching Science in a Cli...* ...as a 1991 paperback, *En el Principio*, published by Libros CLIE of Barcelona, ...ered from T-SELF (P.O. Box 8337, Fort Lauderdale, FL 33310; tel, 800-327-7933) or from the American Scientific Affiliation (P.O. Box 668, Ipswich, MA 01938-0668; tel, 508-356-5656).

A VOICE FOR EVOLUTION *AS SCIENCE*

Background

Science teachers should stress the consistent use of precisely defined scientific terms. Otherwise, students cannot develop an accurate comprehension of scientific knowledge and practice.

Science teachers and scientists concerned about the future of science should (a) recognize the limited scope of science and resist exploitation of science by persons with political, philosophical, or religious agendas; and, while celebrating scientific accomplishments, (b) point out unsolved problems and encourage the investigation of such problems.

In its fiftieth year, the American Scientific Affiliation (ASA) of over two thousand scientifically trained members wishes to go on record in support of the above statements, through an appropriate resolution passed by the ASA Executive Council. As ASA members have explored both their engagement in scientific inquiry and their commitment to the Christian faith, many have sensed problems in the way biological evolution is taught in primary and secondary schools. Noting that at least two major court cases (*McLean v Arkansas Board of Education*, 1982; *Edwards v Aguillard*, 1987) have designated "scientific creationism" (or "creation science") as religious doctrine masquerading as science, the ASA judges it equally important to recognize "evolutionary naturalism" as another essentially religious doctrine masquerading as science. Evolutionary naturalism employs the scientific concept of evolution to promote an atheistic and materialistic view that nature is all there is.

In the current climate of controversy over science teaching in public schools, stretching the term *evolution* beyond its range of scientific usefulness promotes the establishment of evolutionary naturalism. Besides inviting reaction from proponents of scientific creationism, such careless usage also erodes support of sound science education among the broader population of theists, to the detriment of the whole scientific enterprise.

In "The Meanings of Evolution" (*American Scientist*, Vol. 70, pp. 529-31, Sept-Oct 1982) biologist Keith Stewart Thomson identified three commonly employed meanings of the term: (1) the general concept of "change over time"; (2) the hypothesis that all "organisms are related through common ancestry"; (3) a theory setting forth "a particular explanatory mechanism for the pattern and process" described in (1) and (2).

Other meanings range from (4) a scientifically focused concept of populations adapting to changing environments, to (5) a religiously value-laden tenet of naturalistic faith, that "Man is the result of a purposeless and natural process that did not have him in mind" (George Gaylord Simpson, *The Meaning of Evolution*, 1967, p. 345). Science educators should not only distinguish among diverse meanings of evolution but point out that the degree of certainty rightfully associated with them varies widely.

Resolution:
A Voice for Evolution *As Science*

On the basis of the considerations stated above, and after polling the membership on its views, the EXECUTIVE COUNCIL of the AMERICAN SCIENTIFIC AFFILIATION hereby directs the following RESOLUTION to public school teachers, administrators, school boards, and producers of elementary and secondary science textbooks or other educational materials:

BECAUSE it is our common desire to promote excellence and integrity in science education as well as in science; and

BECAUSE it is our common desire to bring to an end wasteful controversy generated by inappropriate entanglement of the scientific concept of evolution with political, philosophical, or religious perspectives;

WE STRONGLY URGE that, in science education, the terms *evolution* and *theory of evolution* should be carefully defined and used in a consistently scientific manner; and

WE FURTHER URGE that, to make classroom instruction more stimulating while guarding it against the intrusion of extra-scientific beliefs, the teaching of *any* scientific subject, including evolutionary biology, should include (1) forceful presentation of well-established scientific data and conclusions; (2) clear distinction between evidence and inference; and (3) candid discussion of unsolved problems and open questions.

(Text of Resolution adopted by the Executive Council of the American Scientific Affiliation, December, 7, 1991)

Teaching Science
in a Climate of Controversy

A View from the American Scientific Affiliation

Committee for Integrity in Science Education
American Scientific Affiliation
P.O. Box 668, Ipswich, MA 01938-0668

The American Scientific Affiliation (ASA) is a fellowship of Christians in the sciences committed to understanding the relationship of science to the Christian faith. Since its founding in 1941 it has grown to a membership of over 2,000. In 1984 the ASA Executive Council authorized formation of a Committee for Integrity in Science Education to address the issues dealt with in this publication. The manuscript was reviewed by the ASA Council and by a number of scientific and educational consultants.

This book was written especially for science teachers and their students. Additional copies of *Teaching Science in a Climate of Controversy* are available as follows:

Quantity	Price
1	$7.00 each
2-9	$6.00 each
10 or more	$5.00 each

Above prices are for pre-paid orders to the same address within the U.S. Non-prepaid orders will include a $1.50/book shipping and handling charge; orders of 5 or more copies will be charged at actual shipping costs only. Order from: *Teaching Science*, American Scientific Affiliation, P.O. Box 668, Ipswich, MA 01938-0668; fax to (508) 356-4375; or call (508) 356-5656. Prices and conditions subject to change.

Executive Council, American Scientific Affiliation, 1986

ISBN: 1-881479-00-5
Library of Congress Catalog Card Number: 86-71822

Committee for Integrity in Science Education, 1986

David Price, *Chair*
Former Science Teacher
Bonita High School, La Verne, California
Springville, California
(Biology, education)

John L. Wiester
Author, *The Genesis Connection*
Buellton, California
(Geology, technology management)

Walter R. Hearn
Adjunct Professor of Science
New College Berkeley
Berkeley, California
(Biochemistry)

Committee Consultants

James O. Buswell III
Dean of Graduate Studies
William Carey University
Pasadena, California
(Anthropology)

Robert B. Fischer
Provost, Senior Vice President
Biola University
La Mirada, California
(Chemistry)

Robert C. Newman
Professor of New Testament
Biblical Theological Seminary
Hatfield, Pennsylvania
(Astrophysics)

Edwin A. Olson
Professor of Geology
Whitworth College
Spokane, Washington
(Geochemistry)

Hugh N. Ross
Former Research Fellow
California Institute of Technology
Sierra Madre, California
(Radioastronomy)

David L. Wilcox
Professor of Biology
Eastern College
St. Davids, Pennsylvania
(Biology)

(Note: Institutional affiliation of persons whose names appear in this book are for identification purposes only.)

To the many scientists and teachers
whose insights are reflected in this publication,
the authors express sincere appreciation.

The American Scientific Affiliation gratefully acknowledges
THE S. G. FOUNDATION
and
THE STEWARDSHIP FOUNDATION
for their generous support and continuing interest.

Additional support for the Third Printing from
THE HENRY PARSONS CROWELL AND SUSAN COLEMAN CROWELL TRUST,
THE R. J. MACLELLAN CHARITABLE TRUST,
and
THE M. J. MURDOCK CHARITABLE TRUST
is gratefully acknowledged.

Additional support for the Fourth Printing from
THE SANTA YNEZ FOUNDATION
and
THE STEWARDSHIP FOUNDATION
is also gratefully acknowledged.

Teaching Science in a Climate of Controversy

CONTENTS

Preface

We had the sky, up there, all speckled with stars, and we used to lay on our backs and look up at them, and discuss about whether they was made, or only just happened—Jim he allowed they was made, but I allowed they just happened; I judged it would have took too long to *make* so many. Jim said the moon could a *laid* them; well, that looked kind of reasonable, so I didn't say nothing against it, because I've seen a frog lay most as many, so of course it could be done.

—Huck Finn, in
Mark Twain's *Huckleberry Finn* (1884)

IN MARK TWAIN'S classic American novel, Huck and Jim wondered about the world and how it came to be. Without knowing much about science or philosophy or religion, they grappled with the profound question of whether *purpose* lay behind the world. Do we owe our existence ultimately to a creator-God or to autonomous forces of happenstance?

All individuals are wise to seek an answer to that question. If the stars "only just happened, " then all of nature can be viewed as being ultimately due to random events and accidental processes. That kind of answer leads some people to see human beings as physical objects devoid of intrinsic worth or value.

On the other hand, if the stars "was made" (in Huck's words), then all of nature was meant to be. We can view ourselves and others as purposeful creatures bearing the stamp of God's intentions. Human life takes on a sacred dimension. We become obligated to treat each other with dignity and respect.

When Huck and Jim decided that the moon gave birth to the stars, however, their answer made no sense at all from a scientific standpoint. No matter how charming and satisfying at the moment, it was contradicted by too much evidence.

Today science teachers find themselves in the middle of an ongoing controversy. The public is divided over what their children should be taught about origins, particularly human origins. Yet when science is properly taught, questions of origins become opportunities for constructive classroom dialogue. Students who learn to distinguish between scientific discoveries and the assumptions buried within them gain a better appreciation of what science is.

For some of the deepest human questions about ultimate meaning and purpose, religious faith is part of the investigative process. The methods of science probe *how* and *when,* but cannot reach "beyond nature" to explore *why* things exist or whether a supreme intelligence is behind our own existence. For many students the two sets of questions appear to be thoroughly entangled.

To help teachers cope with an awkward and complex situation, this booklet was prepared by the Committee for Integrity in Science Education of the American Scientific Affiliation (ASA). ASA wants teachers to present the subjects of origins and of biological evolution with accuracy and openness. Students are better served by valid, up-to-date scientific information than by ideological arguments of strong attackers or defenders of evolution.

The booklet's first section shows how to proceed confidently even in the midst of intense public controversy. The second section discusses four important questions frequently asked by students:

> ▶ **1. Did the universe have a beginning?**
> ▶ **2. Did life on earth arise by chance?**
> ▶ **3. Where did the first animals come from?**
> ▶ **4. What is known of the earliest hominid?**

The third section reminds teachers that they share with scientists a responsibility for public education about science.

In 1984 the National Academy of Sciences (NAS) published a booklet for teachers entitled *Science and Creationism: A View from the National Academy of Sciences.* It provided a broad summary of the evidence on which current scientific conclusions are based, but to some readers, its rejection of "special creation" seemed to imply rejection of a divine Creator. Further, it ignored certain unsolved problems that should be an integral part of scientific education.

Like the NAS publication, the booklet you hold in your hands seeks to protect and enhance the teaching of science in our public schools. It can be used as a supplement to standard biology textbooks. As a member of both NAS and ASA, I commend *Teaching Science in a Climate of Controversy* for its careful treatment of scientific matters and for its practical approach to questions that go beyond science.

John E. Halver
Professor in Nutrition
School of Fisheries
University of Washington
Seattle, WA

I Coping with the Creation/Evolution Controversy

Science can only be created by those who are thoroughly imbued with the aspiration towards truth and understanding. This source of feeling, however, springs from the sphere of religion.

—Physicist Albert Einstein, in
Science, Philosophy and Religion, a symposium (1941)

TEACHING SCIENCE is not easy. Students need to know many things about the world they live in—sometimes more than they want to learn. How the world works is what scientists are continually trying to find out. Their research has led to better materials, more abundant food, faster computers, and so on. But in gaining knowledge scientists have also challenged the way people think, which sometimes causes problems.

In the technologically developed part of the world, the scientific approach has become part of human experience. Other responses to the world—art, poetry, music, worship—are thousands of years older, but today young people without a grasp of science are culturally disadvantaged. And any country without a scientifically literate citizenry will soon find itself economically disadvantaged.

Good science teaching means more than conveying information about what scientists have learned. A more significant task is teaching the particular way scientists look at the world—a way not appreciated by everyone, even in a technologically advanced society.

Valid scientific conclusions are based on valid evidence. Students should learn how to evaluate evidence the way scientists are trained to do. Among other things, that means taking all relevant evidence into consideration while searching for still more evidence.

To teach with openness while upholding standards of scientific integrity does students a great favor. It also contributes to the health of science in our democracy, where basic research depends on continuing support from millions of taxpayers who aren't scientists. What is needed is not blind faith in science but understanding and a reasonable amount of public trust. To retain that trust, science must be taught without omitting important points, overstating its claims, or distorting the truth.

Albert Einstein (1879-1955), one of the world's greatest scientists, recognized that religious understanding has a role to play in all human endeavor, including science.

WARNING: A few critics have faulted this booklet for its incomplete presentation of evidence for evolution. Most high school teachers, however, correctly identify it as a supplement rather than a textbook.

9

The Science Teacher's Dilemma

Americans remain deeply divided in their beliefs about the origin and development of the human species, and a significant number care strongly enough about those beliefs to dispute how to teach the subject in school.

—Attorney and science historian Edward J. Larson, in
Trial and Error: The American Controversy over Evolution and Creation (1985)

BECAUSE SCIENTIFIC knowledge keeps growing, students must always be taught some things that were unknown when their parents went to school. The teacher's role becomes much more difficult if the whole scientific approach seems to run counter to values held by some parents. When that happens, teaching science in the public schools can be a risky business.

At the present time, the teaching of biological evolution in public schools is the subject of a major public controversy. To some extent the controversy arises from simple misunderstandings, but for some it represents a clash of opposing value systems. Science teachers who try to clear up the misunderstandings may feel quite vulnerable to criticism from both sides. In such circumstances, integrity in teaching requires courage as well as wisdom.

Biological evolution is often presented from a historical perspective, beginning with the early nineteenth-century concept of a static creation in which nothing had changed very much "since the beginning." Students are taught how that idea was gradually replaced by a picture of sweeping changes taking place over vast periods of time.

Why did most people eventually abandon the static view of "natural history" held by almost everyone well into the nineteenth century? The answer is that the accumulating geological evidence convinced scientists that the earth had had a very long history—a history of changes far greater than most had thought. None of the present forms of life had been on the earth all that time and many of the older forms had disappeared. Human beings seemed to be relative newcomers to a very ancient planet.

Science and religion were intertwined in the reading of that history in several ways. The static view of the world had been the prevailing way of interpreting both the physical data and the story of creation in the Bible. Yet the biblical picture—of an orderly creation by a dependable God—was one factor that had given impetus to the development of science.

A created world, which God had declared "good," had been deemed worthy of careful study by the pioneers of what was first called "natural philosophy." In a universe that "made sense" because its operations were overseen by a Supreme Intelligence, mathematical description and prediction were both possible and necessary. To quote Galileo, the Bible taught "how to go to heaven" but science described "how the heavens go."

In the two centuries after Galileo, scientists described and predicted many things both in the heavens and on the earth. By the time Charles Darwin's *Origin of Species* appeared on the scene (1859), it was commonly believed that science could eventually explain everything.

Today it is commonly believed that science and biblical religion have always been at war with each other. That belief is not supported by historical investigation. Galileo, for example, was a devout believer in the Bible. Even in Darwin's day, many religious people accepted the evidence for great changes over

Galileo Galilei (1564-1642), like Copernicus, Kepler, and other pioneers of science, was a sincere believer for whom the heavens declared the glory of God.

11

vast periods of geological time—without swerving from a firm faith that such changes were ultimately God's doing. They were able to separate the question of *when* new life-forms appeared from questions about *how* it happened.

The modern picture of how life changed over time was developed largely by geologists who believed in divine creation. The geologic column and the basic facts of fossil succession were established in science (and accepted by most theologians) by about 1840, some twenty years before Darwin proposed a mechanism to explain how such changes had taken place. [1]

Public interest in evolution has always been more complex than a simple clash between two conceptual frameworks, one scientific, the other religious. Further, the science taught in public schools is seldom at the cutting edge of research but rather is what is widely accepted, akin to what legal expert Edward J. Larson calls "public science." The compromise between scientific thought and public policy has its own history, in and out of the courts.

After the famous Scopes trial of 1925, evolution almost disappeared from American high school textbooks. It reappeared only after the Russians launched Sputnik in 1957. Sensing a need to upgrade science education, the federal government funded the Biological Sciences Curriculum Study. The three different high school biology texts produced by BSCS were integrated around evolutionary concepts. In several states, however, earlier laws against teaching evolution in public schools stayed in effect until the late 1960s.

Meanwhile, a movement was taking shape under the leadership of a few individuals with degrees in science whose interpretation of the Bible disallowed acceptance of an ancient earth. They wrote books containing ideas at variance with the main body of scientific thought but conforming to their own religious views. At first they were ignored by the scientific community, which saw nothing new in their criticisms of evolution. But their renewed version of early nineteenth-century science was in accord with the deeply rooted sentiments of many present-day religious fundamentalists.

Organizations sprang up endorsing the modern "young-earth" position known as "scientific creationism." Its followers began assaulting the teaching of evolution in every conceivable way. Activists introduced legislation, testified before state boards of education, threatened lawsuits, ran candidates for school boards, and put pressure on local schools and teachers. In response, small groups of scientists banded together in grassroots political organizations to defend evolution.

Tactics change from time to time. Unable to stop the teaching of evolution, believers in a young earth sought "balanced treatment" for their views, which they claimed were equally scientific. Early in 1982 a federal court ruling struck down an Arkansas balanced treatment act. Litigation over a Louisiana law worked its way up to the U.S. Supreme Court, where in 1987 "scientific creationism" was declared to be a religious view that should not be taught as science.

At the state and local level, political maneuvering continues, frequently focused on textbook selection. One anti-evolution strategy is based on the "establishment of religion" clause in the U.S. Constitution, which prohibits public institutions from favoring one religion over another. The claim is made that Christian children in public schools should be protected from the teaching of evolution. Why? Because evolution, when improperly taught, is seen by many as a doctrine of atheism or "the religion of secular humanism."

Debating whether evolution is "fact" or "theory" often obscures the factual nature of *evidence* and makes scientific theories sound like offhand guesses.

Clarence Darrow (L) and William Jennings Bryan (R) at the 1925 trial of John T. Scopes in Dayton, Tennessee. The American Civil Liberties Union retained Darrow to defend Scopes, a young high school science teacher accused of breaking a state law against teaching evolution in state-supported schools. The conviction obtained by the prosecution, led by Bryan, was reversed on a technicality in 1927, but the law stayed on the books until 1967. In the 1980s, state laws about teaching evolution and "scientific creationism" have been challenged in the courts. One case, over a Louisiana law mandating "balanced treatment," was settled by the U.S. Supreme Court on June 19, 1987. The Court ruled that "scientific creationism" is a religious view, not a legitimate part of science.

As an overall "explanatory principle," evolution remains a broad scientific and philosophical *inference*.

It has been well established (1) that the fossil record shows a succession of of life forms, and (2) that mutation and natural selection provide a plausible mechanism for the formation of new species (sometimes called *micro*-evolution). Other lines of evidence to be considered include the geographic distribution of plants and animals; similarities of "homologous" structures (human arm, bat wing, whale flipper); similarities in embryonic developmental patterns; the genetic makeup of populations; and now structural similarities in the genetic material itself. From the cumulated evidence biologists have inferred a general *macro*-evolutionary principle: "the genetic relatedness of all living things."

Many aspects of evolution are currently being studied by scientists who hold varying degrees of belief or disbelief in God. No matter how those investigations turn out, most scientists agree that a "creation science" based on an earth only a few thousand years old provides no theoretical basis sound enough to serve as a reasonable alternative.

Clearly, it is difficult to teach evolution—or even to avoid teaching it—without stepping into a controversy loaded with all kinds of implications: scientific, religious, philosophical, educational, political, and legal. Dogmatists at either extreme who insist that theirs is the only tenable position tend to make both sides seem unattractive.

Many intelligent people, however, who accept the evidence for an earth billions of years old and recognize that life-forms have changed drastically over much of that time, also take the Bible seriously and worship God as their Creator. Some (but not all) who affirm creation on religious grounds are able to envision *macro*-evolution as a possible explanation of how God has created new life-forms.

In other words, a broad middle ground exists in which creation and evolution are not seen as antagonists. With that middle ground in mind, a teacher need not "take sides" at all.

Some Classroom Guidelines

The extremity of creationist charges and claims is, to a degree, a reflection of corresponding extremities on the part of evolutionists themselves. Both extremes tend to fuel the fires of the other, and to find their worst fears realized.

—Conrad Hyers, in
The Meaning of Creation (1984)

MOST SCIENTISTS defend evolution because they regard it as a key biological concept. Probably most American citizens cherish creation as a basic biblical doctrine. Evolution and creation are often presented as polar opposites, so that if one interpretation holds, the other cannot. In a science classroom, head-on conflict is likely to erupt during almost any discussion of origins. When you as a teacher find yourself in that difficult situation, how should you respond? Here are some suggestions:

1. Use the opportunity. Although an *argument* over evolution with a know-it-all student can be counterproductive, a well-informed teacher can rescue the situation, turning it into a *discussion* that becomes a rewarding experience. Even students who come to class with blind spots in their view of biology can be shown how to ask critical questions, weigh probabilities, and separate facts from opinions. At the same time they can learn to recognize some of their own biases and those of others. If it doesn't get out of control, an energetic controversy can stir up greater interest in science as a whole.

2. Define the limits of discussion. Even the mention of evolution can trigger all sorts of images and issues in some people's minds. When old battles are fought again, old wounds are easily reopened. For more than a century,

evolution has been debated among scientists, philosophers, and theologians. They continue to wrestle with such questions as: How does evolution work? What does it explain? What does it fail to explain? What does it predict? Are there any reasonable alternatives? Could some as-yet-undiscovered mechanisms, biological laws, or rare cosmic events be integrated into the theory to make it more comprehensive in the future? How valid are evolutionary extrapolations from biology into other areas of thought?

Meanwhile, as we have seen, exactly what should be taught about evolution in public schools has become a legal question with political overtones. The subject is debated in legislatures and courtrooms, with media coverage often distorting the issues by focusing on extremes.

With so much going on at so many levels, it is hard to keep classroom discussions on a single level of responsible discourse. While you're trying to explore the limits of biological diversity at one level, someone may interject a biblical reference to "created kinds." The interpretation of scripture then gets mixed up with the interpretation of observations and experiments. Your task is to keep the discussion from roaming so widely that nothing is accomplished. Try to narrow the focus to a few clearly defined questions, but without ridiculing students who do introduce extraneous ideas.

3. Show respect for opposing views. Showing respect for a view one disagrees with means taking it seriously enough to try to put yourself in a proponent's shoes. Concentrate on the logical steps taken as students with different viewpoints interpret various kinds of evidence. Try to avoid making assumptions about motives, which are easily misjudged. The logical weight of an argument is what counts in scientific discussions, and a teacher should model the way a scientist might approach an unsolved problem. Typically, scientists pick apart their own ideas first, trying to anticipate any arguments that might be raised by others when their ideas are made public.

4. Consider the whole spectrum of opinion. Advocates of extreme positions tend to paint a win-or-lose, either/or picture. What some see as a con-

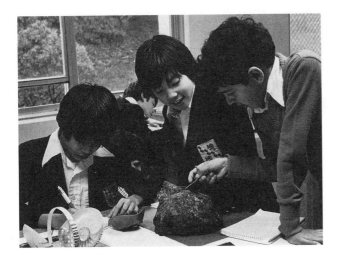

test between real science and a dangerous pseudoscience, others see as a defense of real religion against a blasphemous belief in "godless evolution" or "mere chance." Yet between those two extremes lies that broad middle ground where real science can coexist with real faith in God.

Portraying a disagreement as a clash between two warring camps is a common device for simplifying important issues. But to fortify one intellectual position against all others, and treat them as a single opposing force, is generally unrealistic. The cultivation of moderating positions is likely to be more productive but requires more effort. In fact even recognizing the existence of a moderating position demands the rethinking of one's own position. Nevertheless, continual rethinking is an important part of the way scientists operate.

5. Seek common ground. Suppose you find that class discussions of origins quickly gravitate toward polarizing positions. Then what do you do? It might be worthwhile to take up each position and let its adherents begin to work it out in detail. If that can be done without making students feel defensive, a kind of turning point may be reached—starting from either direction.

At such a point the "airtight logic" of an extreme position begins to break down, giving way to recognition that all is not so tidy. Thus, students who call themselves creationists may suddenly see that "creation scientists" cannot claim much knowledge of *how* God has created anything. Or they may see that *when* creative acts took place is still an open question among biblical scholars (Psalm 90, for example, which is attributed to Moses, speaks of differences in divine and human ways of reckoning time).

On the other hand, students who consider themselves evolutionists may see that explaining how things have *developed* may not tell us much about how they *began*. They may come to recognize that for any major transformation proposed on the basis of the fossil record, the mechanisms are far from understood. And even if we fully understood such mechanisms scientifically, the philosophical and religious questions of direction, design, meaning, and purpose would remain—beyond science.

In the end, proponents on both sides may come to agree on at least one (generally correct) point: it is easier to advocate a general position than it is to support it by detailed argument. One might say that extremists who oppose each other on nearly everything else do agree on one (generally incorrect) point. Both argue that no middle ground exists.

6. Watch your language. Terminology is a problem. For example, although *chance* has important philosophical implications in some usages, to a chemist a "chance collision of atoms" refers merely to a reaction that can be analyzed statistically. The word *sudden* may mean within a time-span of a microsecond to a physicist, of a few minutes to a bacterial geneticist, or of several hundred thousand years to a geologist.

The terms *evolution* and *creation* themselves cause much confusion. Astronomers use the term *evolution* to refer to the aging of stars and galaxies. In biology, it can mean small-scale changes within species, genera, and families; it can also mean a total change as large as from amoeba to human. Making such distinctions clear may be the most important classroom contribution of a well-informed science teacher.

Sometimes the scientific term *evolution* takes on the connotation of a world view recognizing *chance* (in the sense of happenstance or accident) as a kind of elemental driving force. Such a conviction, called by some *evolutionism*,

can function as a pseudoreligion. When *evolution* and *chance* are used in that sense, they could appropriately be spelled with a capital *E* or *C* (the way *God* is written with a capital *G*).

The word *creation* also has a broad range of meanings. Theists (including Christians, Jews, and Muslims) believe in God as both creator and sustainer of the universe; they view natural law as a reflection of God's wisdom and power. Atheists, on the other hand, believe that no creator exists and that natural law is autonomous and self-existent. Agnostics take no firm position on God's existence or on the "nature" of natural law. In between, many nineteenth-century scientists were deists, believing that God played no further role in the universe beyond initially winding it up like a clock.

All theists are creationists in that broad sense, and this includes many who are scientists with professional credentials and years of research experience. Where the scientific data are inconclusive, theists may express reservations about macro-evolution, especially since the biblical narrative seems to emphasize certain stages of divine creation. Hence such terms as *special creation, progressive creation,* and *theistic evolution* have been coined. In the heat of debate and in much popular writing, qualifying adjectives tend to drop off, leaving the erroneous impression that all creationists are united against all evolutionists.

7. Keep asking questions. Some students demand pat answers but most will appreciate openness. It is unnecessary (and in many circumstances unwise) for a teacher to "take sides" in class on the religious issue of Creator versus no-creator. But asking the right questions can help students identify such issues and distinguish science from *scientisms*—the philosophical positions that claim to be verified by science.

Many statements, whether made by scientists or theologians, cry out for someone to ask: What are you talking about? What do you mean? What's the evidence? How do you know? What assumptions are you making? What other conclusions make just as much sense—or more sense?

You might try some of these discussion-starters with your students, then develop your own examples:

—A leader in the creation science movement writes that "the truth of creation and the myth of evolution continue to be recognized by more and more people everywhere."

—Reporting on a cat-sized fossil primate from 18 million years ago, a popular science magazine concludes, "Other than the fact that it was one of the earliest apes on man's evolutionary line, little is known about *Proconsul africanus.*"

—A television personality introduces a science program with the solemn words, "The Cosmos is all there is, there was, or ever will be."

—A Nobel prize-winning physicist includes in his book on the early history of the universe the statement that "the more the universe seems comprehensible, the more it also seems pointless."

—A newspaper account of a lecture on sexuality in primates reports an anthropologist's view that divorce, adultery, and promiscuity are inborn traits stemming from man's evolutionary past, "some 5 million years ago, when man came down out of the trees" (headline: PROMISCUOUS? BLAME IT ON APE FOREBEAR").

—An evangelist says of a fossil orangutan 16 million years old: "It is impossible to prove the age of anything concerning which we do not have reliable, intelligent, verbal, eye-witnesses."

17

Correcting Past Mistakes

We must bring science back into life as a human enterprise, an enterprise that has at its core the uncertainty, the flexibility, the subjectivity, the sweet unreasonableness, the dependence upon creativity and faith which permit it, when properly understood, to take its place as a friendly and understanding companion to all the rest of life.

—Scientist-statesman Warren Weaver, in
Science and Imagination (1967)

ONE OF THE encouraging things about science is that it tends to correct itself. Scientists expect to be challenged about how they fit all the pieces together. When writing for other scientists, they tend to express their conclusions rather cautiously. They know that their ideas might be overturned by more evidence or by a more complete understanding of the evidence already at hand. In fact, they try to find the weak points in their own arguments before anyone else does.

In popular writings about science, however, caution is sometimes thrown to the wind to make conclusions sound more substantial or dramatic. In a public controversy, things may get even worse. Advocates of a particular side, becoming more dogmatic and defensive, often spend most of their effort exposing weak arguments of the *other* side.

One way for a teacher to show the human side of science is to point to mistakes made in the past but corrected by further work. As evolutionary science has developed, mistakes have in fact occurred. In the current climate of controversy it might be helpful to take a similar approach toward "creation science." Two stories that provide a kind of balance are those of Piltdown Man and the Paluxy River "man-tracks."

The two stories have many parallels. Neither piece of evidence was central to the main arguments of either side. Doubts about the validity of each were expressed by scholars on both sides from the beginning. More was made of the evidence in the popular press than in scholarly discussion. Eventually, after careful investigation, the data were declared defective and the conclusions drawn from them invalid. Looking back, both stories seem embarrassingly trivial, yet arguments based on both lines of evidence lasted for about the same time period—about forty years.

Piltdown Man. Ever since Charles Darwin published *The Descent of Man* (1871), paleoanthropologists (scientists who study human ancestors) have searched for "the missing link" between humans and other primates. From the late nineteenth century to early in our century the search was concentrated in Europe, where most scientists lived. In France and Spain some exquisite cave paintings were attributed to prehistoric Cro-Magnon Man, an early representative of our species known through abundant skeletal remains.

In Germany, France, and Belgium, skeletons of the prehistoric Neanderthal Man were discovered. They were measurably different from modern human skeletons in certain features, especially in skull shape. Actually, the Neanderthal brain cavity was larger, on the average, than that of modern humans.

Assuming the validity of Darwin's hypothesis that humans had descended from ape-like ancestors, some scientists expected to discover a transitional form having a human-size brain but ape-like characteristics in the rest of the skull, especially in the jaws and teeth. Ape jaws are more rectangular than human jaws, and the front side teeth are more dog-like. The "brain-first" idea was that an ancestor evolving toward humanity would have first developed intelligence, as shown by a larger brain case; human-like jaws and teeth and an upright walking posture would have developed later.

In 1912, Charles Dawson and zoologist A. Smith Woodward announced to the Geological Society of London the discovery of an early human skull fragment accompanied by an ape-like lower jaw. Found side by side in a gravel pit near Piltdown in County Kent, England, the two fossils were suggested to be from a single skeleton, a human ancestor dubbed "Piltdown Man." From the beginning, though, some odd features of the discovery site were recognized. In particular, animal fossils of various ages were found in close association. Those problems were not enough to quench the general enthusiasm over finally having found a true forerunner of the human species.

The story of how eagerly Piltdown Man was accepted by most of the British scientific establishment has often been told. But why did Piltdown remain a member of our ancestral hierarchy for so long (forty-one years) in spite of growing evidence against it? For one thing, the theory of evolution was under such constant attack by religious leaders and public figures that dogmatic defenders of evolution tended to grasp at any evidence suggesting human evolution. Also, the fact that the discovery was made in England seemed to bolster British national pride, both in their scientists and their own presumed ancestry. Finally, Piltdown fit the "brain-first" theory prevailing at the time; it looked exactly like the expected missing link.

Some scientists never did believe that the ape-like jaw and teeth were from the same skeleton as the human-like skull. Their doubts were strengthened by fossil discoveries in Java, China, and especially South Africa. Finally, in 1953, Piltdown was revealed to be a fraud—one of the great hoaxes of modern history.

The unmasking of the Piltdown forgery is an excellent example of the self-correcting character of science. By a nonradiometric dating technique (comparing the content of the element fluorine with that in other fossils) the skull fragment and jaw were shown to be of modern age, much younger than some other fossils found in the same gravel.

The teeth in the lower jaw were then re-examined by a suspicious dentist, who observed that someone had filed them down. Further, the fragments had been chemically stained to match the color of the gravel and to give them an appearance of great age. Clearly an unknown forger had "salted" the deposit with a human cranial fragment and the lower jaw of an orangutan, both cleverly doctored to fit contemporary expectations.

To this day, the forger's identity has not been revealed. Most investigators of the case consider the scientists who discovered the "fossils" to have been innocent victims of the hoax, although the perpetrator must have been scientifically informed. Nonetheless, the discoverers, along with the many others

A 1915 painting by John Cooke of some of the principals in the Piltdown controversy. Sir Arthur Keith is shown examining his reconstruction of the skull from pieces discovered by the two men standing to his left, lawyer Charles Dawson (L) and zoologist Arthur Smith Woodward (R). Forty years after its discovery, Piltdown was shown to be a fake.

who accepted the find as genuine, were guilty at least of unwarranted confidence in their own objectivity.

The Piltdown hoax stands as a reminder that scientific judgments can be influenced by pressure of pride and career, by the culture, personality, and even nationality of scientists. It teaches all of us an important lesson: preconceived ideas, even reasonable scientific ideas, can strongly influence our interpretation. Sometimes we see what we are looking for rather than what is really there.

Footprint IIS+4 (under shallow water) from the Taylor trail in the Paluxy riverbed near Glen Rose, Texas. Once considered human by some observers, the 24-inch elongated track has shallow relief patterns and rust-brown color distinctions indicating dinosaurian digits.

The Paluxy River Man-Tracks. In the 1930s a paleontologist from the American Museum of Natural History in New York, Roland T. Bird, was startled by what he saw in a New Mexico store window. Two large pieces of limestone on display each showed what looked like a perfect but oversized human footprint. The scientist was told that they came from a particular site in central Texas, where dinosaur-like tracks could also be seen.

Near the town of Glen Rose, Texas, Bird found many dinosaur tracks in the bed of the Paluxy River. He uncovered many more by digging through one limestone layer and a few inches of clay to get to a thicker limestone layer bearing the tracks. He cut out several big blocks containing trackways and shipped them on flatcars back to New York, where they are still on display in the American Museum's hall of dinosaurs.

In his 1939 report in *Natural History*, Bird did not mention finding any human tracks in the Paluxy river bed. According to old-timers in the area, during the depression years local people not only excavated tracks to sell but also carved fake tracks in pieces of limestone. In 1950 an article by Clifford L. Burdick about the Paluxy tracks was published in a Seventh Day Adventist journal, *Signs of the Times*. Burdick's photographs of oversized human-like tracks purportedly cut from the river bed appeared in *The Genesis Flood* (1961), by John C. Whitcomb, Jr., and Henry M. Morris, one of the books that led to the modern young-earth creationist movement.

According to the accepted geologic timetable, dinosaurs became extinct some 65 million years ago, tens of millions of years before humans appeared on earth. Since occurrence of human footprints in the same strata as dinosaur prints would be devastating to that timetable, accounts of the story appeared in a flood of anti-evolution publications. A film called "Footprints in Stone" (made at the site), which argued for a recently created earth, was shown in many churches.

Careful investigations have failed to confirm the human character of any of the prints. A 1975 paper by Berney Neufeld entitled "Dinosaur Tracks and Giant Men" in another Seventh Day Adventist journal, *Origins*, showed photographs of three-toed dinosaur tracks and trails of elongated tracks with no human features. The same article contained photographs of cross-sections cut through several museum specimens supposedly taken from the Paluxy River site. A genuine dinosaur track showed compaction beneath the footprint, but a "human" track, which appeared to be carved, did not.

Neufeld warned creationists that "the Glen Rose region of the Paluxy River does not provide good evidence for the past existence of giant men. Nor does it provide evidence for the co-existence of such men (or other large mammals) and the giant dinosaurs."

In 1980 a young biology graduate named Glen J. Kuban made the first of many trips from Ohio to Texas to study the tracks. Measuring the "Taylor" tracks, still described as human in some creationist literature, he noted many

20

nonhuman characteristics. Between 1982 and 1984, Texas high school teacher Ronnie J. Hastings, a Ph.D. in physics, also concluded that none of the tracks was human. He, geologist Steven Schafersman, paleontologist Laurie Godfrey, and anthropologist John Cole dubbed themselves "Raiders of the Lost Tracks." They made a videotape, "Footprints in the Mind," later called "The Case of the Texas Footprints."

In 1984 Kuban and Hastings began working together at the site. In 1985, studies of the Paluxy River footprints filled an entire issue of *Creation/Evolution* (a small journal "dedicated to promoting evolutionary science"). Papers by the "Raiders" and other scientists, with photographs and detailed drawings, offered several plausible explanations for the alleged man-tracks, some of which were shown to be imperfect tracks of dinosaurs. Kuban has written a monograph, *The Texas Man Track Controversy*, and published some of his own photographs and drawings in the Spring/Summer 1986 issue of *Origins Research* (a newspaper providing "a means for students and educators to critically analyze the evolution and creation models of origins").

Kuban's correspondence and on-site conferences with certain creationist leaders have led some to reevaluate their position. Notably, the producers of "Footprints in Stone" have withdrawn that film from further distribution. Even more significant was the January 1986 about-face of one of the most widely read young-earth publications, *Acts & Facts* of the Institute for Creation Research (ICR). A feature article by John M. Morris, son of ICR director Henry M. Morris, concluded that "it would now be improper for creationists to continue to use the Paluxy data as evidence against evolution."

This example and that of Piltdown Man provide glimpses of how science works. Perhaps reviewing past mistakes on both sides and seeing how they were corrected will help students become more openminded, whatever position they—or their parents—take on human evolution.

High overhead view of tracks at the Taylor site, Paluxy riverbed in Texas. A trail of deep three-toed dinosaur tracks is crossed from the lower right by elongated tracks. Until recently those tracks were claimed to be human by some opponents of the accepted geologic time-scale. Evidently the tracks were made by a bipedal dinosaur walking in plantigrade fashion, placing weight on its soles and heels.

II   Teaching about Origins: Four Open Questions

A revitalization of interest in scientific honesty and integrity could have an enormous benefit both to science and to the society we serve.

> —Physicist Lewis M. Branscomb, Chief Scientist of the IBM Corporation, in "Integrity in Science," *American Scientist* (Sept.-Oct. 1985)

WHEN SCIENCE is presented as an open-ended but self-correcting human endeavor, students gain appreciation for unanswered questions within science and for questions that move beyond science into philosophy and religion.

Dealing with questions of origins constructively helps students clarify the issues in their own minds, and can increase the general public's understanding and support of science. But the present climate of controversy tempts each side to play up the ideological biases of the other. Hence some creationists may argue that "in a divinely created world, an intrinsically atheistic premise must lead to false conclusions." Some evolutionists may argue that "creation science" cannot possibly be good science because "it stems from a religious commitment."

In science it doesn't really matter where ideas come from. What matters is the kind of *evidence* that supports those ideas—a point that needs to be stressed in science classes. In each subsection that follows, an important question of origins is discussed, with emphasis on the present state of the scientific evidence on which conclusions must be based. Each discussion also makes at least one specific point about the nature of scientific conclusions. Look for the following points:

1. Science continually raises philosophical questions that go beyond the competence or purview of science.

2. Evidence of random chemical processes is not necessarily evidence for philosophical accidentalism.

3. In science, an unanswered question is far more important than an unquestioned answer.

4. In science, tentative conclusions should be stated in tentative form.

5. The confidence expressed in any scientific conclusion should be directly proportional to the quantity and quality of evidence for that conclusion.

(Opposite page) The great Andromeda galaxy, M31 (NGC 224), nearest of the spiral galaxies, approximately 2 million light years away from us. The two other largest spots in the photograph are other galaxies. Other spots of light are some of the 200 million stars in our own galaxy, the Milky Way, along the telescope's line of sight to Andromeda.

Did the Universe Have a Beginning?

What is the ultimate solution to the origin of the Universe? The answers provided by the astronomers are disconcerting and remarkable. Most remarkable of all is the fact that in science, as in the Bible, the world begins with an act of creation.

—Astronomer Robert Jastrow, In
Until the Sun Dies (1977)

Albert Einstein and Edwin Hubble at the Mount Wilson Telescope.

HUMAN BEINGS HAVE probably always wanted an answer to the question, "Where do we come from?" Scientists have kept reaching further back to include the origin of the earth and now even the origin of the entire universe. For Aristotle the universe was eternal and had no beginning. In our own day astronomer Fred Hoyle has echoed the same note with an attempt to avoid "arbitrary starting conditions." Although the idea of an eternal universe has satisfied many thinkers, modern discoveries have made that idea more difficult to accept. Today the best scientific evidence points to a real beginning, not only of the matter and energy of the universe but of time and space as well.

An Expanding Universe. The first hint that the universe might be expanding came from an accidental discovery reported in 1914. While looking for something else, astronomer Vesto Slipher noted that certain spiral nebulae were moving away from our earth and sun at enormous speeds. Grasping the importance of that observation, Edwin Hubble and Milton Humason trained their giant Mt. Wilson telescope on those nebulae, now called galaxies. Between 1925 and 1930 Hubble and Humason measured the recessional speeds and distances of enough galaxies to show that they are moving away from our galaxy at speeds proportional to their distances from us.

The great nebula in Orion, M42, a turbulent mass of gas so large that it takes light six years to cross it. The Orion nebula is a region of active star formation by accretion of gas and dust from the explosions of other stars.

24

If all galaxies are moving away from us (and from each other), the whole universe must be expanding. We can visualize this by watching a pattern of spots spread out on the surface of a balloon as it is inflated. That picture, though inadequate, gives us some idea of what the astrophysicists mean by an expanding universe. (More precisely, they say that our universe is actually expanding into four-dimensional "space-time.")

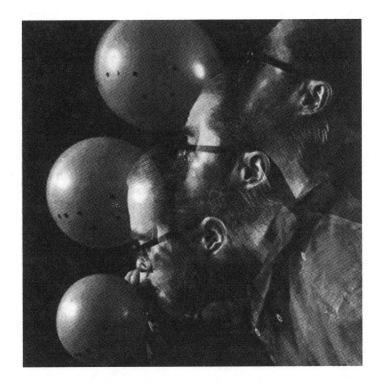

Imagine going back in time, when the galaxies were closer together than they are now. The further back we go, the closer they must have been, so we can imagine getting to a point where all the known galaxies were compressed together in a very small volume. Equations of relativity theory have been experimentally verified to a high enough precision to describe the behavior of the universe [2]. They indicate that compression could become so great that the universe would be a point of no size at all and thus of infinite density. Matter and energy as we know them would not exist, and space and time would have no meaning.

A universe-in-a-point is admittedly almost beyond human imagination; scientists call it a singularity, an absolutely unique event. It would correspond to the beginning of the universe, or at least to a time at which no more meaningful information could be obtained. Thus the scientific evidence of an expanding universe points to a universe with a beginning.

The "Death" of Stars. Other evidence of a beginning comes from the chemical composition of stars. By studying the light coming from them, we have learned that stars consist mainly of hydrogen, the lightest element. We now know that hydrogen is used up during a star's "lifetime." Within the nuclear furnace at the center of the star, hydrogen is continuously converted to helium and other heavier elements. Just as a campfire converts wood to ash and then dies out when the wood is used up, so stars move toward "death" as they use up their hydrogen.

The largest stars end that process in a catastrophic explosion called a supernova, which scatters into space the heavier elements produced inside the star. Evidently the planets of our solar system were formed from such debris, which is why astronomers have spoken of the earth as made of "the ashes of stars." Because the change from hydrogen to heavier elements is irreversible, the amount of hydrogen in the universe is constantly decreasing. The universe thus resembles a clock that is running down.

Although the death of stars, like the expanding universe, points to a beginning, some scientists have persisted in seeking an eternal universe. In 1948 three British astronomers, Hermann Bondi, Thomas Gold, and Fred Hoyle, proposed a new model for such a universe, which came to be known as the *steady-state universe* or *continuous creation* model. They suggested that new matter was constantly being created in the form of hydrogen atoms or neutrons. As the galaxies moved apart, they said, new galaxies formed from the new matter and occupied the vacancies created by expansion. How was the new matter created? Hoyle proposed an unknown force called the C-field (for "creation field") which supposedly created the new matter out of nothing.

Creation *ex nihilo* ("out of nothing") seemed to those scientists to be simply a law of nature. For some years the continuous creation hypothesis was regarded as a strong competitor to the idea that the universe had a beginning. The major problem with continuous creation was that no observations supported it. For example, no trace of newly appearing or extinct galaxies has been detected. Further, if true, it would have violated most conservation laws in physics, although the amount of violation in each instance would probably have been too small to detect with existing instruments. At any rate, the debate continued until the mid 1960s, when new evidence came to light.

The Cosmic Background Radiation. In 1965 two scientists at Bell Laboratories were annoyed by "static" in a powerful new microwave radio receiver they were trying to use. Arno Penzias and Robert Wilson first thought they had located the trouble when they found pigeons nesting in the huge, horn-shaped antenna. Evicting the pigeons did not get rid of the so-called static. They continued their search for its source, and eventually found it, a discovery that earned them a Nobel prize in physics.

Penzias and Wilson found that the mysterious microwave radiation came from beyond their receiver, from beyond the earth, and even from beyond our galaxy. The whole universe seemed to be emitting a faint "glow" of microwave radiation wherever they pointed their antenna. What they had discovered, now called the cosmic background radiation (not to be confused with cosmic rays), appears to be the diluted remnant of the fierce heat and light emitted in the early moments of a primordial explosion.

The background radiation can be likened to the heat and light given off by the dying coals of a fire. Red-hot coals tell us that the fire was burning quite recently. As time passes, the coals become dimmer and duller in color. From the time the fire is flaming until the coals lose their color, we can see the radiation as visible light. Even after the coals have stopped glowing, we can

Arno A. Penzias (L) and Robert W. Wilson (R) in front of the microwave antenna with which they discovered the cosmic background radiation. The two scientists shared the 1978 Nobel Prize in physics for their discovery, which provides strong evidence that the universe began in a "Big Bang."

still feel heat radiating when we put our hands close to the ashes. That radiation is no longer in the visible but in the infrared region of the spectrum.

One could say that the "ashes" of the universe's primal "fireball" are by now very, very cold and no longer emit even infrared radiation. Instead, they emit the less energetic longer-wave radiation called microwaves, detectable by certain sensitive radio receivers. Like all radiation, microwaves exist in the form of "light particles" called photons; the background-radiation photons correspond to the very low temperature of only three degrees above absolute zero.

Amazingly, almost twenty years before its discovery, physicist George Gamow had predicted the existence of such background radiation as a consequence of his "hot" model of the universe. [3] Using Gamow's model, Ralph Alpher and Robert Herman predicted in 1948 that the cooling off of the initially very hot universe would by now yield a background radiation corresponding to a temperature roughly five degrees above absolute zero. [4] Today the presence of this universal background of microwave radiation convinces most scientists that the universe not only had a beginning but began in the gigantic explosion now called "the Big Bang."

Expansion at a Decreasing Rate. Another piece of evidence supporting the Big Bang was discovered by Allan Sandage of the Mt. Wilson and Palomar observatories. After many detailed observations and calculations, he reported in 1974 that the galaxies are moving away from each other at decreasing rates of speed. The observed deceleration of the galaxies is one more indication that, like a clock that was once wound up, the universe had a beginning.

Could the Universe Be Oscillating? Some scientists continue to search for evidence that the universe is eternal. A model proposed by Ernst Opik suggested that the Big Bang was actually a "Big Bounce," with the universe expanding and contracting like an accordion. In Opik's view the universe would complete one cycle of expansion and contraction every hundred billion years or so. Those attracted to the idea of an oscillating universe include such science popularizers as Carl Sagan and Isaac Asimov. It is a view that requires no beginning.

Recently, however, it has been demonstrated that even if the universe contains enough mass so that its gravity eventually halts the present expansion and forces a collapse, the collapse will *not* produce a bounce. [5,6] The term *entropy* is used for the energy of a system that is unavailable to perform work; because of the huge entropy generated by the universe, any ultimate collapse would be more like a "crunch" than a bounce. The universe, it seems, either expands forever or goes through only one cycle of expansion and contraction.

Before the evidence accumulated in support of the Big Bang, Fred Hoyle said he saw no philosophical grounds for preferring a cosmic beginning. He considered it a "distinctly unsatisfactory notion," since it put the basic assumption out of sight where it could never be challenged by a direct appeal to observation. Such an outlook reminded him of how primitive people try to explain things, "obliged by their ignorance of the laws of physics to have recourse to arbitrary starting conditions. These are given credence by postulating the existence of gods." [7]

Did Hoyle's view change after the Big Bang model became widely accepted? Not really. In a 1978 interview he said that the main controversy is whether the "so-called origin of the universe" really has to be taken literally or is a physical transition from a preceding state. He had "little doubt that there has to be a preceding stage, perhaps even an evolutionary process." [8]

Hoyle's tenacious commitment to an eternal, self-existent universe appears to stem more from a philosophical bias than from empirical evidence. Astronomer Robert Jastrow has suggested that scientists' reluctance to acknowledge a beginning for the physical universe originates in the hope that science has the capacity to explain everything.

Diagrams illustrating three possible models for the universe. Currently most evidence points to the validity of the "Big Bang" model.

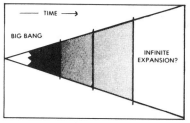

(a) BIG BANG

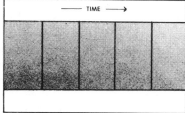

(b) STEADY STATE

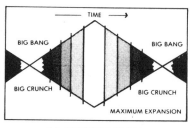

(c) OSCILLATION

Going Beyond Science. With all the current scientific evidence indicating that our universe had a definite beginning, we are led to ask many other questions: Where did the universe come from? What existed before it began? What was the source of the incredible power manifesting itself in the Big Bang's blaze of light?

Since there is little likelihood of obtaining concrete evidence from "beyond the Big Bang," most scientists probably agree with geologist Preston Cloud (in his *Cosmos, Earth and Man*, 1978) that such questions "transcend the bounds of science."

Some physicists, such as Alan Guth, continue to search for a new physical theory that could explain the origin of the universe *ex nihilo* in accord with the concepts of quantum mechanics. The vacuum from which Guth's universe theoretically arises, however, is not a true vacuum; it contains energy. Attempts to derive a truly *ex nihilo* origin for the universe by what is known as "quantum tunneling" have so far been frustrated. Quantum mechanics places severe limitations on "virtual particle production," and general relativity places demanding limits on the origin of time and space.

Even if such a theory could be developed, we would be left with some fundamental questions: Why is there *something* rather than *nothing*? Are the forces of nature really self-existent or were they preconceived? How can we account for the design of such elegant patterns capable of creating our vast and complex universe out of nothing?

Ultimate questions like those are approached from a philosophical perspective, not a scientific one. We need not be surprised, nor feel embarrassed, when scientific discovery leads us into such questions. Of course, those who expect too much of science may be disappointed to learn that science has inherent limitations.

People need to understand that *science continually raises philosophical questions that go beyond the competence or purview of science.*

Did Life on Earth Arise by Chance?

Scientific research on the origin of life is in an exploratory phase, and all its conclusions are tentative.

*—Science and Creationism: A View from
the National Academy of Sciences* (1984)

IN SCIENCE one question generally leads to another. For example, if the universe had a beginning, then life in the universe must have had a beginning. Did life on earth arise from the chance collision of atoms and molecules? Today almost all high school biology textbooks that pose this question state or imply that it did.

Life as a Chemical System. In contrast to studies of the heavens, the study of life as a chemical system is quite recent. But plants and animals, unlike the stars, existed here on earth where scientists could "take them apart" to study them. That study has paid off. Today, in fact, biochemistry (or "molecular biology") is one of the most rapidly developing experimental sciences.

Over a century ago the term *organic* designated the chemistry of substances produced by living organisms. Those substances were recognized as different from the "inorganic" components of rocks and minerals. At first that difference was attributed to some kind of "vital force" residing in living things. (Today organic chemistry refers basically to the chemistry of carbon compounds, whether synthesized in the laboratory or isolated from natural sources.)

Plants and animals turned out to be composed of ordinary elements—chiefly carbon, hydrogen, oxygen, nitrogen, phosphorus, and sulfur. Those elements, however, have been found to bind together to form large polymer molecules of almost unlimited variety. Today the chemistry of life deals largely with the interactions of complex biopolymer molecules.

A simple one-celled bacterium may contain thousands of such intricately structured macromolecules. Among them are the nucleic acids, composed of smaller building blocks (nucleotides) strung together in a definite, non-random sequence. One type of nucleic acid, *DNA* (the "double helix"), carries within its structure what is now called the "genetic code." In its nucleotide sequence lies the information for synthesizing all the macromolecules in that cell. Another type of nucleic acid, *RNA*, helps to translate the code into the manufacture of important macromolecules called proteins.

A cell can be thought of as a biochemical factory. One group of proteins makes up the walls dividing the factory into compartments. Other proteins called enzymes serve as specific "machine tools" to carry out each step in the manufacturing process. DNA in the cell nucleus provides the "blueprints" for converting raw materials (nutrients taken in by the cell) into final products. When the final product is a duplicate of the whole factory, we have a simple picture of how cells reproduce.

Such an analogy, while hinting at what goes on in a living cell, enormously oversimplifies biology at the molecular level. Translation of the genetic code, for example, takes place at cellular structures known as ribosomes, themselves composed of perhaps fifty different proteins and a number of different RNA molecules. Indeed, calling something as intricately programmed as RNA or

DNA a "molecule" is like calling an important novel "a bunch of words." We are talking about *huge* molecules bearing enormous amounts of structural information.

On the borderline between the nonliving and the living are the viruses. Much smaller and simpler than cells, they have to borrow some machinery in order to reproduce. They get it from a living cell which they infect. Carrying their own blueprints, though, the viruses are also composed of complex nucleic acid and protein molecules. The tiniest bit of proto-life is thus assembled from millions of atoms arranged in a very precise way. The leap in complexity from the chemistry of minerals to that of the very simplest living thing is therefore immense.

Bridging the Gap. In a letter written in 1871, Charles Darwin expressed serious doubt that life could ever arise today "in some warm little pond." He reasoned that any protein molecule formed by spontaneous chemical interaction would quickly be consumed by already living things. A few years before, in France, Louis Pasteur had effectively refuted claims that the *spontaneous generation* of life had ever actually been observed. By showing experimentally that life did not arise in sterilized media, Pasteur also laid the groundwork for modern microbiology, the study of the simplest living organisms.

Are today's scientists less skeptical about that possibility, now called *abiogenesis* or *chemical evolution*? Yes and no. *Yes* because it is recognized that conditions on the primitive earth were different from conditions today. At some stage there were no plants to enrich the atmosphere in oxygen. In a reducing (non-oxidizing) environment the spontaneous synthesis and persistence of organic precursor molecules might have occurred. Scenarios proposed by a number of scientists depict the accumulation of such compounds in surface waters to form an organic "soup," from which life is thought to have emerged. Typical scenarios are outlined on the opposite page.

The hypothesis that life arose by stepwise chemical processes in no way requires optimum conditions everywhere on our planet. Suitable conditions may have been very specialized and localized, perhaps utilizing absorption on clay in a microclimate from which oxygen was somehow excluded. Since scientists are visualizing unknown steps in an overall process of low probability, Darwin's "warm little pond" will do as well, for the time being, as a whole ocean of organic soup.

One relatively new piece of information has been the discovery of many very simple organic compounds deep in interstellar space where there is no suggestion of life. Closer to home is evidence from laboratory experiments in which very simple compounds—generally reduced gases such as methane, ammonia, and hydrogen cyanide—are allowed to react under various conditions in an oxygen-free aqueous environment. Many different mixtures have been tested with a variety of energy sources from heat to ultraviolet radiation. In the original experiment reported by Stanley Miller in 1953 the gases were passed through an electrical discharge. The organic compounds produced in such experiments include most of the basic building blocks (monomers) for the biopolymers—that is, for the proteins and nucleic acids.

On the *no* side of abiogenesis is recognition that producing the monomers, as impressive as that was thirty years ago, is a long way from producing a cell, a virus, or even a biologically active protein or polynucleotide. Further, new geological evidence has cast doubt on the concept of a strongly reducing early atmosphere on earth [9]; such an atmosphere was assumed in designing

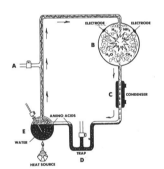

Simplified drawing of the type of apparatus used by Stanley Miller in 1953 and others since then to synthesize organic compounds from mixtures of gases thought to have been present in the Earth's early atmosphere. Gases introduced at *A* are subjected to a spark discharge or other source of energy at *B*. A heat source at *E* boils water, which is condensed at *C*. Products accumulate in aqueous solution at *D*. Even if such experiments produce appropriate monomers under plausible conditions, investigators are still a long way from producing even the simplest form of life, as shown in the chart on opposite page.

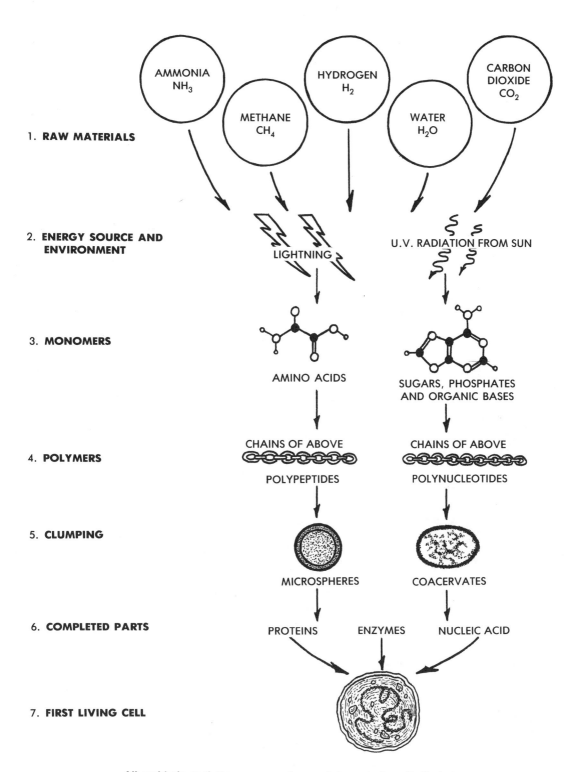

1. **RAW MATERIALS**

AMMONIA
NH_3

METHANE
CH_4

HYDROGEN
H_2

WATER
H_2O

CARBON
DIOXIDE
CO_2

2. **ENERGY SOURCE AND
ENVIRONMENT**

LIGHTNING

U.V. RADIATION FROM SUN

3. **MONOMERS**

AMINO ACIDS

SUGARS, PHOSPHATES
AND ORGANIC BASES

4. **POLYMERS**

CHAINS OF ABOVE

POLYPEPTIDES

CHAINS OF ABOVE

POLYNUCLEOTIDES

5. **CLUMPING**

MICROSPHERES

COACERVATES

6. **COMPLETED PARTS**

PROTEINS   ENZYMES   NUCLEIC ACID

7. **FIRST LIVING CELL**

All prebiotic evolutionary scenarios contain *many* hypothetical steps.

ISSOL-86, the 8th International Conference and 5th Meeting of the International Society for the Study of the Origin of Life, drew 250 scientists from 27 nations to the U. of California at Berkeley in July 1986. Chemist Clifford N. Matthews of the U. of Illinois, Chicago (top), argued that HCN in a nonaqueous environment could have led to both polypeptides and polynucleotides. Chemist A.G. Cairns-Smith of the U. of Glasgow in Scotland (bottom) argued that life began with "crystal genes" of clay minerals, later added proteins and nucleic acids.

the first prebiotic synthesis experiments. In addition, NASA's 1976 Viking missions to Mars failed to detect any evidence of life, or of appropriate prebiotic organic compounds in the Martian soil. It seems that life does not appear automatically even on a planet with a number of important similarities to earth [10].

Some methane and ammonia may have reached the earth through collisions with comets, but the conclusion of many geologists and atmospheric physicists is that the early earth's atmosphere consisted primarily of nitrogen, water vapor, and carbon dioxide. That makes an enormous difference in evaluating experiments in prebiotic synthesis. Energetically, making amino acids from hydrogen, methane, and ammonia is like letting water run downhill; trying to make the same amino acids from nitrogen and carbon dioxide is like trying to force water to run uphill.

Although much chemical information has been obtained in laboratory experiments, the goal of producing life is nowhere in sight. In fact, there is no assurance that scientists are even on the right track. Like NASA's Apollo missions, the early prebiotic synthesis experiments were a great achievement, something never done before. But those successes were rather like reaching the moon when one's goal is Alpha Centauri, the nearest star. The energy and engineering it took to reach the moon, three earth-days and 240,000 miles away, are entirely different from what would be required to reach Alpha Centauri, 25 trillion miles away.

Will we ever reach the stars? No one knows, but at present, few space scientists would bet on it. Will we ever synthesize life? That is not clear either, but certainly no one should object to probing life's best-kept secrets. At the same time it is proper to call attention to the dimensions of our ignorance.

At present we know of no energy-coupling system not derived from a living organism that can convert simple monomers into the energy-rich, information-laden biopolymers necessary for life. Some kind of selection is necessary to drive the reactions in that direction. We are faced with a paradox: "Both nucleic acids and proteins are required to function before selection can act at present, and yet the origin of this association is too improbable to have occurred without selection." [11]

Scientists are now pursuing reports of catalytic activity in certain RNA molecules and studying undersea thermal vents as model environments. Yet a veteran researcher insists that there is simply no evidence that an appropriate mixture "rearranged itself at random until a living cell emerged." [12]

Discovery of some directive process that could have functioned at the dawn of life would possibly change the situation, but today the assembly of monomers into highly ordered biopolymers by random processes must be considered *extremely* improbable. Some scientists think the event would still be extremely improbable even if billions of planets had been covered with solutions of the monomers for billions of years. [13] At this stage of our scientific knowledge, it would be irresponsible to give students the impression that "life arose by chance." Scientists do not know how life arose.

Can A "Chance" Experiment Ever Be Performed? Like the origin of the universe, the origin of life is a scientific question that leads beyond science. For physicists, chemists, and molecular biologists, terms such as *accident, chance,* and *randomness* properly apply to collisions between atoms and molecules. Even though nothing can be said about the behavior of an individual

32

atom or molecule, with millions of them under study the overall probability of such a collision occurring can be accurately calculated.

When biochemists use an expression like "random collision" or "chance event," they are saying nothing at all about any underlying *cause* of such an event—provided they are speaking scientifically. If they are speaking philosophically, or even popularly, they could be stating a personal belief about the ultimate cause of events.

Suppose that one day biochemists do set up conditions in which simple compounds react to form complex molecules, and those molecules aggregate into cell-like structures having catalytic activity. Would that successful experiment demonstrate what "chance" can accomplish in driving chemicals toward life?

At one level, it would. But it would also make clear that chance had little or nothing to do with setting up the highly specific conditions for such an experiment. You can be sure that afterward the experimenters would not say, "Oh, that experiment was an accident; it just happened." No, they had a goal in mind and used their intelligence to achieve that goal. They would rightly claim the credit for what happened. So, we might ask, if the element of purpose is always there to "contaminate" experiments intended to demonstrate *pure chance*, is chance ever "pure"?

We assert, therefore, that the methods of science are quite limited for discerning whether or not some ultimate purpose or design resides behind what happens in the physical world. Of course many individuals, on philosophical or religious grounds, are convinced that an underlying purpose exists. Their opposites, who might be called accidentalists, argue that chance is all there is. At the scientific level, however, to say that something happened randomly or by chance says nothing at all about whether it happened *on purpose* at another level.

Whatever we read in the newspapers, or in scientific journals, about life's origins, we should remember that *evidence of random chemical processes is not necessarily evidence for philosophical accidentalism.*

First photograph ever taken on the surface of the planet Mars, a few minutes after the Viking 1 spacecraft landed in July 1976. One of the spacecraft's footpads appears in the lower right corner. Viking 2 landed on Mars in September 1976. NASA scientists were excited when photographs of frost on the rocky Mars landscape reached earth, indicating the presence of at least a trace of water, but no evidence of life or of prebiotic organic compounds has been obtained. In the entire universe, only our planet Earth is known to support life.

Where Did the First Animals Come From?

It is important to note that no major pattern of scientific evidence that conflicts with Darwin's theory has turned up.
—Biology (Scott, Foresman textbook, 1988)

The evidence we find in the geologic record is not nearly as compatible with Darwinian natural selection as we would like it to be.
—Curator of geology David M. Raup, in "Conflicts Between Darwin and Paleontology," *Field Museum of Natural History Bulletin* (Jan/Feb 1979)

Life's History on Earth

TO EXPERIENCE SCIENCE as a process of discovery, students need to encounter some real unsolved problems. Science is not for those who merely want to "learn the facts."

Evolutionary biology is no exception, despite a public controversy that pushes beleaguered biologists to stress evolution as *fact*. The two extremes in the public mind seem to be either that (1) "evolution did not occur" (that is, life has not changed significantly), or (2) "evolution has explained everything."

Evolutionary *theory* is concerned with figuring out what specific changes occurred, and by what mechanisms. A taste of real biology comes from exploring what is left to discover about life's history as well as what is already known. Opportunity to examine both aspects is provided by the question, Where did the first animals come from?

According to evolutionary theory, modern animals arose from earlier forms, and those from still earlier ones. Theoretically a continuous trail should lead back to an original life-form. How complete is the trail of evidence found in the fossil record?

For over a century, paleontologists have searched for the intermediate or transitional forms expected from Darwinian theory. The "reptilian bird," *Archeopteryx*, is an example of an exciting discovery that seems to fit the general picture, whether or not it was an actual ancestor of modern birds. In following the succession of fossilized life from ancient layers of rock to more recent ones, we see that, in general, the number of diverse life-forms has increased over time.

On the other hand, the fossil record has remained remarkably silent about what went on between the world of single-celled protozoa and that of the first real animals, the marine invertebrates. The search for evidence all over the world has turned up new and exotic forms and has increased our general knowledge of past biological communities. But it has not yet defined the steps by which the animal life known today originated.

The Known. What do we know about the earliest animals? Certain types of multi-celled animals left their mark in late Precambrian rocks about 700 MYA (million years ago). Impressions found between sedimentary layers define some simple animals that had neither shells nor skeletons. Those earliest fossilized animals were first discovered in Australia about twenty-five years ago. Called the Ediacara fossil complex, they have subsequently been found in South Africa, England, the Soviet Union, and Newfoundland.

Dating back more than 100 million years before the Cambrian period, the Ediacaran ecosystem is well preserved within sandy sediments. [14] Some

34

of its fossils look as if they might be related to later jellyfish, segmented worms, echinoderms, or corals, although most are strange-looking creatures unlike later animals. Are any of these flat, soft-bodied animals ancestral to modern phyla? No one is really sure. (A *phylum* is a major biological category or *taxon*, characterized by a unique structural plan called *Bauplan* in German; *phyla, taxa,* and *Bauplaene* are the plural forms.)

Beginning with the Cambrian period (570 to 500 MYA), enormous numbers of marine animal fossils are found. For that reason the interval from the beginning of the Cambrian to the present is called the Phanerozoic Eon—which means the period of manifest or obvious life. Some paleontologists estimate that as many as 100 new body plans appeared during the early Cambrian period. Of that number, only some 30 phyla remain today, the rest having become extinct. [15] Most major invertebrate classes within the phyla also came into existence then. Trilobites were among the many classes that later became extinct.

The Unknown. Is it possible to give a mechanistic account of the origin of the marine invertebrate phyla? To call their dramatic appearance at the beginning of the Phanerozoic an example of evolution is like giving the problem a name rather than solving it. A huge "morphological distance" separates the protozoans from the early Cambrian fauna and even from the Ediacaran assemblage. A whole series of hypothetical transformations would be required to form organisms as complex as crustaceans or snails, which have many specialized types of cells organized into unique tissues and organs.

Despite other successes, biologists understand very little about the profound changes that took place somewhere near the Precambrian/Cambrian boundary. Almost no fossil evidence exists prior to the Cambrian to show any definite relationships between phyla. Perhaps the missing data will yet be discovered in the extensive Precambrian sedimentary deposits underlying Cambrian formations throughout the world. The fossils of the Ediacaran period show that the late Precambrian seas did contain many creatures, but with little obvious relationship to the phyla that replaced them.

Many biologists are confident that the "neo-Darwinian synthesis" will eventually account for such phylum-level divergence by the accumulation of tiny changes. Others remain open to new mechanisms because the evidence does not seem to fit the Darwinian pattern. As paleontologists Douglas Erwin, James Valentine, and John Sepkoski [16] describe the situation,

> The fossil record suggests that the major pulse of diversification of phyla occurs before that of classes, classes before that of orders, and orders before families. This is not to say that each higher taxon originated before species (each phylum, class, or order contained at least one species, genus, family, etc., upon appearance), but the higher taxa do not seem to have diverged through an accumulation of lower taxa.

The Possibilities. Figures A and B on p. 36 illustrate hypothetical evolutionary patterns for the appearance of new phyla. Phyletic gradualism (A) is the classic neo-Darwinian prediction of continuous, slowly developing morphological change in branching lineages. In the major rival theory of punctuated equilibria (B), periods of morphological *stasis* are interrupted by rapid morphological change emerging from speciation in small, isolated populations. In both models morphological distance is achieved through a cumulative process based on natural selection.

The problem is that neither A nor B fits the emergence pattern of the animal phyla observed in the fossil record, which is more like Fig. C. Essentially

TRILOBITE -- *Olenellus eagerensis* -

SCLERITES OF *Chancelloria*,
EARLY CAMBRIAN, SIBERIAN PLATFORM

(Above) Some Early Cambrian fossils.

Diagramatic Representation of How New Animal Phyla Emerge

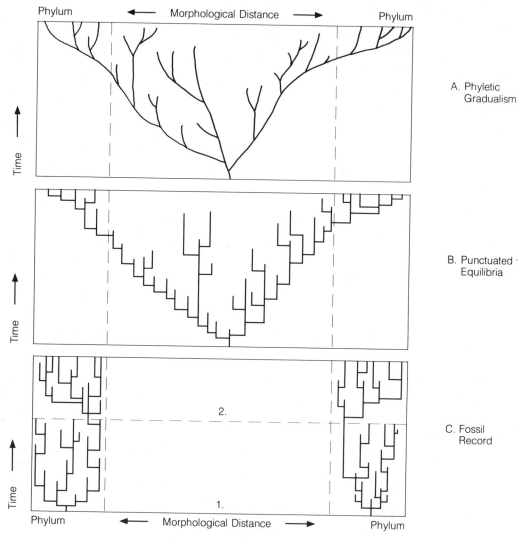

Phylum ← Morphological Distance → Phylum

A. Phyletic Gradualism

B. Punctuated Equilibria

C. Fossil Record

Time

Phylum ← Morphological Distance → Phylum

Patterns of emergence of new animal phyla, drawn with arbitrary scales of time and morphological distance and with only a few species represented. **Phyletic Gradualism (A)** is the neo-Darwinian picture of continuous change over the lifetime of essentially all species. **Punctuated Equilibria (B)** depicts periods of *stasis* or stability at the species level, interspersed with (geologically) sudden morphological changes that arise from isolated speciation events. In both "bottom-up" models, evolutionary novelty emerges bit by bit, eventually achieving enough morphological distance to merit higher-level taxonomic distinctions as new orders, classes, and phyla.

The fossil record (C) shows a kind of "top-down" pattern, with essentially all extant and extinct animal phyla emerging near the base of the Cambrian, some 570 MYA (as at 1, Fig. C). That is, the phyla seem to appear first, followed by lower and lower taxa in a burst of species diversification. After a mass extinction (as at 2, Fig. C) such as the Permian of 248 MYA, no new major body plans (phyla) appear, but the pattern again resembles rapid variation on already existing themes, filling the newly available adaptive space (See R. Lewin, "A Lopsided Look at Evolution," *Science,* Vol. 241, 15 July 1988, p. 291-3). Mass extinctions may have prevented Darwinian processes from establishing the long-term pattern of life's history (see S. J. Gould, "The Wheel of Fortune and the Wedge of Progress," *Natural History,* March 1989, pp. 14-21; "Tires to Sandals," *Natural History,* April 1989, pp. 8-15).

all the known animal phyla (*Bryozoa* being a possible exception) originated in a period of perhaps two percent of life's tenure on earth.

That is why the base of the Cambrian has been called a "crux in life's history," when "nearly all basic designs of invertebrate life entered the fossil record for the first time." [17] Those designs became established within a few tens of millions of years, and "since then life has again and again consisted of variations on a theme." [18]

In the "top-down" pattern pointed out by Erwin, Valentine, and Sepkoski, the relatively sudden appearance of a new Bauplan is followed by a burst of diversification based on that plan, radiating to fill the available adaptive space. Mass extinctions are followed by periods of new evolutionary/ecological opportunity. Such periods show very high levels of lower taxon diversification, but it is almost entirely variation on already established themes. Only early in the Cambrian did large numbers of new phyla and classes appear.

Although various explanations have been offered for the observed pattern, none is fully satisfactory. One ecological hypothesis suggests that only the Cambrian world could have presented such "adaptive emptiness," in which almost any mutant morphology could survive. A rival "genomic" hypothesis speculates that in the one-time-only transformation from single-celled existence, drastic genetic changes may have occurred, possibly through some kind of "horizontal transfer" of genetic material. [19] It remains true that "The most dramatic kinds of evolutionary novelty, major innovations, are among the least understood components of the evolutionary process." [20]

At present, the mechanisms to explain a "megapunctuational event" like the origin of the invertebrate phyla are largely hypothetical. Many biologists seem to be saying that because higher taxonomic categories have their own properties, innovation at one level may not be explainable by simple extrapolation from a lower level. Are genetic mechanisms that operate at the species level (such as natural selection acting on point mutations) adequate to account for the larger pattern of life's history? That question is still open [21]. So is the question of what factors have prevented the formation of new phyla over the past 500 million years. The primary effect of natural selection at higher levels may be to produce stasis—to inhibit rather than promote major evolutionary change.

Meanwhile, it is inappropriate to insist that the origins of higher taxa have been explained by known microevolutionary mechanisms. With so much left to be done, responsible teachers of evolution will emphasize unsolved problems as well as past accomplishments. Some student who gets excited about the subject today may one day help explain how major body plans originated, as exemplified in the Cambrian explosion.

In science, an unanswered question is far more important than an unquestioned answer.

Cliff in central East Greenland rising about 1,700 m (5,600 ft). It is part of a Late Precambrian sedimentary rock formation about 15 km (over 9 miles) thick.

What Is Known of the Earliest Hominid?

When considering our origins it is clear that we have often been less than objective.

—Anthropologist Richard Leakey, in
The Making of Mankind (1981)

EVOLUTIONARY THEORY holds that humans and apes evolved from a common ancestor. Historically, it was similarities in anatomy that first suggested a close relationship between humans (hominids) and certain apes (pongids). Today, claims of "common descent" are increasingly based on structural similarities in the information-bearing molecules (proteins, DNA, or RNA) of living primate species.

On a 1989 NOVA television program, anthropologist Vincent Sarich stated that human DNA and chimpanzee DNA differ by only about 2 percent. That tiny difference, he said, could nevertheless represent 20 million mutations. Then he posed a rhetorical question: "If humans were created, why were we created in the image of a chimpanzee?" [22] It was the kind of theological ("Why?") question that goes beyond science, but there are many scientific questions that might be asked instead. For example, in what sequence, and in what sorts of populations, did those mutations occur? Presumably they began in some primate species at a "branch point" where the hominid line would have diverged from a line leading to chimps.

Any proposed *phylogeny* or "family tree" begins with an *inference* of ancestry. (Similar reasoning is used to argue that existing biblical manuscripts must have been copied from earlier but as yet undiscovered ones.) An outpouring of new molecular data is intensifying old debates among geneticists and paleontologists over the order and lengths of various "branches." In higher primate investigations, for example, it is not yet clear how well the pieces of evidence from various "molecular clocks" will support each other. [23]

Paleoanthropologist Richard Leakey and African coworker study an *Australopithecus* fossil site near Lake Turkana in Kenya.

The Search for a Common Ancestor. In *Bones of Contention*, Roger Lewin points out that paleoanthropology shares with other historical sciences "the limitations of trying to reconstruct events that happen only once: there are no experiments to be done that can confirm or deny the major themes that are sought." [24] Although some progress has been made in tracing our most recent ancestors, the picture gets more complex and the task more difficult as we go further back in time.

Evidence of the earliest hominid has been sought ever since Darwin published *The Descent of Man* (1871), but no fossil yet discovered is generally recognized as representing that particular "missing link."

Over the years, many candidates have been proposed as links in a chain leading *from* that hypothetical ancestor to *Homo sapiens*. Some have represented false starts in the human parade; others have been overshadowed by new discoveries. A few have gained enhanced recognition through additional discoveries—as when Richard Leakey discovered a more complete skeleton of *Homo erectus* in Africa. Overall, the scarcity and incompleteness of fossil specimens have been major problems in drawing a clear picture of the human family tree.

The attention of most paleoanthropologists is now focused on Africa. Ancient primate fossils found there, especially those of *Australopithecus africanus* ("South African ape"), have ape-like skulls but skeletal features suggesting upright posture. It was therefore theorized that our human ancestors came down from the trees in response to environmental pressures of shrinking forests and spreading grasslands, and thus learned to walk upright.

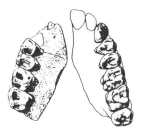

Possibly much later, it is thought, a higher intelligence evolved, as evidenced by the increasing size of the brain cavity in the more recent fossil specimens, *Homo habilis* and *Homo erectus*. Today that "feet-first" or "brain-later" hypothesis is the accepted theory of human evolution, guiding anthropologists as they propose fossil lineages.

In the 1960s and '70s, the fossil finds of the alleged ape-like ancestors of human beings were generally arranged in the following sequence:

$$Ramapithecus \longrightarrow Australopithecus\ africanus \longrightarrow Homo$$

By the early 1980s, *Australopithecus africanus* had been replaced by *Australopithecus afarensis,* the most famous specimen being "Lucy," discovered by Donald Johanson. *Ramapithecus* ("Rama Valley ape") had disappeared completely from the human family tree. This deletion of what was acclaimed in 1964 as "the oldest probable forerunner of man" has not been widely publicized. Nor is it widely known that the role of *Ramapithecus* as our earliest known human ancestor had been challenged by a number of anthropologists from the beginning. Somehow the creature found its way into many textbooks without mention of such reservations.

In fact, a 1980 college biology textbook assured students that our roots "began with a woodland ape 12 million years ago." It identified that ape as *Ramapithecus,* presumed to be the first of our fossil ancestors to diverge from the common human-ape (hominoid) line. [25]

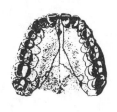

Despite widespread scientific and popular acceptance of *"Rama"* as a human ancestor, its status was based on extremely scanty evidence. Full-body reconstructions were made on the basis of (1) seed-eating behavior (deduced from the way the teeth were worn) and (2) several pieces of jawbone reassembled to resemble the parabolic curve of a human jaw. The re-creations showed a man-like ape walking uprightly [See Richard Leakey's *The Making of Mankind* (1981) and *Human Origins* (1982)].

Upper jaws of *Ramapithecus* (top), orangutan (center), and human (bottom). The outline of the *Ramapithecus* jaw has been superimposed on the other two for comparison.

Paleontologist David Pilbeam at the rim of the Ngorongoro crater in Tanzania, Africa.

With the finding of pieces of the cranium and skeleton in Pakistan and China, interest in *Ramapithecus* has continued, but the creature is no longer considered a human ancestor. Now it is regarded as a stage in the lineage of the orangutan. [26]

The brightest spot in the *Ramapithecus* story is that it illustrates the willingness of scientists to correct their own mistakes. David Pilbeam (formerly at Yale, now at Harvard) was one of the original proponents of *Rama* as a human ancestor. On his return from Pakistan in 1979 after finding new fossil evidence, Pilbeam readily admitted his earlier misinterpretation of the evidence. [27]

Unsolved Problems. A college textbook of physical anthropology summarizes the current situation in early hominid evolution:

> Though it is unfortunate that a clear and uncomplicated picture of early hominid evolution cannot yet be presented, we fervently hope that future discoveries will provide evidence to bring the scenario into focus. [28]

The emphasis in that statement, as in all good science, is on the need for *evidence*. While we await the evidence needed to bring that hypothetical scenario into focus, our thinking about human origins should not ignore the following points:

1. *The last ancestor assumed to be common to both apes and humans has not yet been found.* Current biochemical studies of the DNA "molecular clock" indicate that the supposed common ancestor would have lived 5 to 8 million years ago (MYA). At present we have no known fossil hominoids of *any* kind from that particular period.

2. *There are problems in putting together lineages leading to the modern apes as well as to man.* Reassignment of *Ramapithecus* from human to orangutan ancestry leaves the orangutan with probable fossil ancestors dating from 14 to 9 million years ago (MYA). The orangutan trail picks up again with fossil teeth found in caves on the mainland of Asia dating at 1.5 MYA. Tracing the fossil ancestry of chimpanzees and gorillas has been more difficult. Miocene proto-apes left a relatively abundant fossil record in Africa from 23 to 15 MYA. Although a "tree ape" (*Dryopithecus*) was found in Europe dating from about 13.5 to 10 MYA, there is a gap of 11 million years in the African fossil record until 4 MYA when the australopithecines first appear in Africa. But the australopithecines (from 4 to 1.5 MYA) are currently assigned to lineages related to or leading to *Homo*.

In other words, there is no fossil record of the stem or branch of the ancestral tree leading to the chimpanzee or gorilla, which are the living apes whose DNA sequences most closely resemble our own. Why should those modern apes lack fossil ancestors?

One explanation offered is that the humid conditions of tropical forests were not conducive to the formation or preservation of fossils. Yet studies of fossil vegetation tend to confirm the presence of a tropical forest environment at many sites where Miocene proto-apes have been recovered. It is also possible that too much uncertainty exists to assign hominoid fossils to the correct ancestral lineages.

3. *The fossil remains of the australopithecines may have no direct relationship to the human line.* The specimen called Lucy, dated at between 3 and 4 MYA, has received much publicity, but it is not yet clear how she fits into the scheme. Lucy was an adult about four feet tall. She and her related family might belong in our ancestral line, but with the exception of a human-like pelvis, she seems to have had few, if any, human characteristics. [29]

Richard Leakey's best guess now places all of the australopithecines in an extinct category outside the line of human ancestry. [30] If his suggestion is

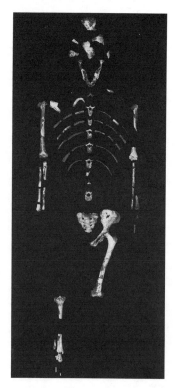

(Below) "Lucy," a female skeleton of *Australopithecus afarensis* from Hadar, Ethiopia, is 40 percent complete. Found by Donald Johanson, the skeleton has been dated at 3 million years.

correct, we have found no connecting links at all until 2 MYA, when the genus *Homo* (as presently classified) first appears in the fossil record.

4. *Strong pressures seem to have existed against the evolution of walking on the hind feet, the so-called bipedal locomotion.* Most mammals use all four limbs, giving them efficient movement and greater running speed. Recent studies of fossil carnivores from the African grasslands indicate that many kinds of fast-running hunters and scavengers existed during the times thought to be significant for bipedal human evolution, making slow movers more likely "dinners" than "diners." One proposed advantage of bipedalism under those circumstances is the ability to carry food in the hands. Another is the ability to see over tall grass, but at present the whole subject is highly speculative. It has even been suggested that behavioral-adaptive reversals could have been common in primate lineages; that is, one species might have come down from the trees and another returned to the trees.

5. *No satisfactory mechanistic explanation has been offered for the development of the large human brain.* This is striking in view of its creative potential—as illustrated by the production of cave paintings, symphonies, and (?) booklets on the so-called creation-evolution controversy. It is difficult to visualize the kind of unexploited ecological niche in which these and other uniquely human capabilities would have had survival value.

Better reproductive strategies are usually stressed as the driving force behind evolutionary advances. Yet the large head-size of human infants relative to the mother's small pelvic structure seems to go against that line of argument. Human beings appear to be the mammals with the greatest difficulty in childbirth. That problem is generally ignored in textbook discussions of human evolution.

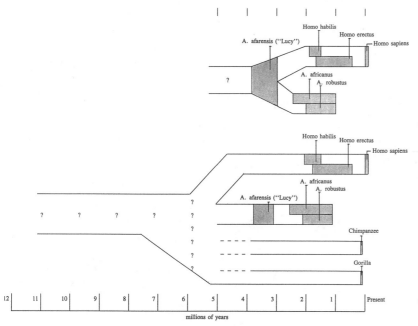

Current status of the missing link. A standard interpretation is shown below, with an alternative interpretation in which *Australopithecus afarensis* ("Lucy") is postulated as a link, above. There are no recognized African ape fossils for the entire 12-million-year period. Interpretations of hominid evolution have been complicated by skull WT 17000, dated at 2.5 million years. The remarkably complete cranium, discovered at Richard Leakey's Lake Turkana site in Kenya by Alan Walker of Johns Hopkins, is causing "a major rethinking of human origins." (A. Walker, *et al.*, "2.5 Myr *Australopithecus boisei* from west of Lake Turkana, Kenya," *Nature*, Vol. 322, 7 Aug 1986, p. 517.)

6. *A very large theoretical superstructure supporting human evolution has been built on a comparatively small data base.* In *The Making of Mankind*, Richard Leakey quoted David Pilbeam as saying, "If you brought in a smart scientist from another discipline and showed him the meager evidence we've got, he'd surely say, 'Forget it, there isn't enough to go on.' " Leakey commented that, of course, the anthropologists couldn't *take* such advice, "but we remain fully aware of the dangers of drawing conclusions from evidence that is so incomplete." The specialists not only have a relatively small number of distinct fossil types to work with but also disagree on how to classify the few fossil types they do have.

A Balanced Assessment. In conclusion, too many problems remain unresolved and too many pieces of evidence are missing to say that the search for human origins is over. "The origin of the Hominidae [the family to which humans belong] is still a scientific enigma and accurately determining it is a fascinating pursuit." [31] That pursuit must continue, but in view of the past record of discredited missing links and the temptation to make up for lack of key evidence by overstatement, a cautious and humble attitude is surely the best policy in the sensitive area of human origins.

In *The Natural History of the Primates* (1985), the eminent primatologists John and Prune Napier conclude their section on the origin of apes and man with this comment: "Much new and better material will have to be recovered before the history of the various groups can be told with any degree of confidence." [32]

In the National Academy of Sciences booklet *Science and Creationism* (1984), a four-paragraph section on Human Evolution cites strong molecular evidence that humans, chimpanzees, and gorillas are closely related. The first paragraph discusses fossils linking humans to "ancestral primates," asserting that "the 'missing links' that troubled Darwin and his followers are no longer missing." [33] The third paragraph links *Homo sapiens* to "our earliest ancestors" by a "succession of well-documented intermediate forms or species," [34] leaving the impression that the branching of hominids from other primates is well documented in the fossil record. In fact such documentation is far from complete. The lessons of Piltdown and *Ramapithecus* remind us that *in science, tentative conclusions should be stated in tentative form.*

Science has a bearing on our understanding of what it means to be human, but science is not the only aspect of human endeavor with an important stake in that question. In all areas of scientific inquiry and perhaps especially when human origins are under consideration, *the confidence expressed in any scientific conclusion should be directly proportional to the quantity and quality of evidence for that conclusion.*

Paleoanthropologist Donald C. Johanson at the Hadar site in Ethiopia, continuing the search for hominid ancestors.

III   Conclusion: Your Role as Teacher

It is absurd to suggest that some things are not better understood than others, or to deny our young people the tools they need to get ahead in a knowledge-intensive world.

—Physicist Lewis M. Branscomb, Chief Scientist of the IBM Corporation, in "The California Decision on Science Textbooks," *American Scientist* (Nov.-Dec. 1985)

SCIENCE TEACHERS uphold the standards of scientific integrity by showing students how to arrive at conclusions based on valid evidence, and by teaching with openness. The unique character and significance of scientific explanation can be presented without ignoring or belittling other approaches to human understanding.

Teachers share with scientists a responsibility for public education about science. The editorial quoted above was written by the president of Sigma Xi, The Scientific Research Society, urging scientists to do a better job of communicating to the public.

Lewis Branscomb wants people to realize that science and ethical or religious values "need not be in conflict." He points to the fact that many scientists are deeply religious, adding, "It is hard to study science and not believe that our existence has some meaning, some purpose outside of ourselves." There is plenty of room for tolerance on both sides, he suggests.

The language of science is the language of probability, not of religious faith nor even of mathematical certainty. As Branscomb shows, scientific understanding is based on mathematical models—but we know that all our models are incomplete, "and doubtless not unique."

These words of caution apply especially to models of origins, where there is seldom even a hint of mathematical precision. Assertions are sometimes made about "what must have happened" or "what could not possibly have happened" in the initial fireball, on the primitive earth, or to produce the first human beings. The absolute quality of such assertions is seldom based on mathematical considerations alone.

Statements about events in the distant past almost always have ideological presumptions imbedded in them. The basic presumption about origins concerns whether a creative purpose was ultimately responsible, or happenstance. Even those who deny the existence of a divine Creator sometimes use phrases like "nature's design" or "creatures of evolution" to describe what has come into being. Scientific understanding of the natural world should augment one's sense of purpose, or at least of wonder and awe.

Lewis Branscomb counsels that the public schools are not the places for ideological battles. To the nation's scientists, he said:

> Let's trust the teachers to be sensitive to students' feelings and family traditions, sharing with our children the wonders of nature as discovered through the tools of scientific exploration.

This booklet asks teachers to honor that trust. As you engage in scientific explorations with your students, recognize that, for many citizens, "the wonders of nature" reveal the glory of God. Let science be science, but let it also be a discipline to enlarge their vision of the Creator.

It is equally important, of course, to be sensitive to those who do not share that vision. You have a challenging task—to present science as an open-ended discipline leading to a better understanding of the natural world.

Helen E. Martin, teacher of earth and space science at Unionville High School in Pennsylvania, exemplifies the kind of integrity, openness, and creativity this booklet seeks to encourage. After she and her students built and operated their own weather-satellite tracking station, she was named the 1987 International Lecturer for the National Science Teachers Association (U.S.A.) and the Association for Science Education (U.K.). Ms. Martin serves on the Board of Directors of the National Board for Professional Teaching Standards, established to improve teaching in the United States. The National Board was proposed in *A Nation Prepared: Teachers for the 21st Century,* a 1986 report of the Carnegie Forum on Education and the Economy. The Task Force that wrote the report was chaired by Lewis M. Branscomb, then IBM's Chief Scientist. Ms. Martin, who has a B.A. in mathematics and an M.Ed. in physical science, has taught math and science at Unionville High School since 1967.

Appendix: Some Additional Resources

THE TECHNICAL JOURNALS in which scientific research papers are published can be consulted in university and many college libraries. Readable reviews of scientific topics are available in many magazines, such as *Science, American Scientist, Scientific American, Science News, The Smithsonian, Natural History, Bioscience, The Science Teacher, The American Biology Teacher,* and (from England) *Nature* and *New Scientist.*

Many practicing scientists have written books for the general public as well as textbooks. *Cosmos, Earth and Man* (Yale Univ. Press, 1978) by geologist Preston Cloud covers most of the scientific topics touched on in this booklet. Among college textbooks, *Evolution* (W. H. Freeman & Co., 1977) by Theodosius Dobzhansky, Francisco J. Ayala, George L. Stebbins, and James W. Valentine is a modern classic.

SCIENCE is most interesting to students when taught as an open-ended enterprise with many questions yet to be answered. Critiques of "the general theory of evolution" include *The Great Evolution Mystery* (Harper & Row, 1983) by science writer Gordon Rattray Taylor, *Evolution: A Theory in Crisis* (Adler & Adler, 1985) by molecular biologist Michael Denton, and *The New Biology* (New Science Library, 1987) by philosopher Robert Augros and physicist George Stanciu.

On the unsolved problem of "abiogenesis," *Origins: A Skeptic's Guide to the Creation of Life on Earth* (Bantam Books, 1987) by Robert Shapiro, a chemist active in the field, is accurate, up to date, and fun to read. A critical evaluation at a more technical level is *The Mystery of Life's Origin: Reassessing Current Theories* (Philosophical Library, 1984) by Charles Thaxton, Walter L. Bradley, and Roger L. Olsen.

A fascinating look at both the scientists and the science of human paleontology is provided in *Bones of Contention: Controversies in the Search for Human Origins* (Simon and Schuster, 1987) by Roger Lewin, editor of research news for *Science* magazine. Lewin's beautifully illustrated *In the Age of Mankind* (Smithsonian Books, 1988) presents the current scientific picture of human origins as a mixture of exciting discoveries and perplexing questions that remain to be answered.

DOZENS OF recent books deal specifically with the so-called creation-evolution controversy. Young-earth creationists have poured out a steady stream of books attacking evolutionary science and the geologic timetable. Mainstream scientists have responded with books for the general public strongly defending evolution. Between polemic extremes are some excellent books by well-informed writers. Several inexpensive booklets are also available to help teachers deal with "scientific creationism" in the classroom. *Science and Creationism: A View from the National Academy of Sciences* (National Academy Press, 1984) was prepared by a Committee on Science and Creationism of the most prestigious scientific organization in the United States. *Modern Science and the Book of Genesis* (National Science Teachers Association, 1986), written by James W. Skehan, a geologist who is also a Jesuit theologian, includes NSTA's excellent 1985 Position Statement on "Inclusion of Nonscience Tenets in Science Instruction."

WHEN STUDENTS bring up religious questions concerning creation, science teachers can get help from several books that bring together the writings of scientists and religionists. One useful collection, *Is God a Creationist? The Religious Case Against Creation-Science* (Scribners, 1983), was edited by humanities scholar Roland Mushat Frye. Another, *Evolution and Creation* (Univ. of Notre Dame Press, 1985), was edited by philosopher Ernan McMullin; it has an excellent chapter by geneticist Francisco Ayala outlining "The Theory of Evolution: Recent Successes and Challenges." Both books include excellent introductory chapters written by the editor.

Robert W. Hanson also wrote an excellent introduction on "Science or Belief, A False Dichotomy," for the collection he edited, *Science and Creation: Geological, Theological, and Educational Perspectives* (Macmillan, 1986). Based on a 1982 AAAS symposium, that book also contains the complete texts of the 1981 Arkansas "Balanced Treatment" Act 590 and of U.S. District Court Judge William R. Overton's decision in the *McLean v. Arkansas Board of Education* trial over that law. Another scholarly symposium, *God and Nature: Historical Essays on the Encounter between Christianity and Science* (Univ. of California Press, 1986), edited by historians David C. Lindberg and Ronald L. Numbers, contains Numbers's informative chapter describing "The Creationists."

STUDENTS from conservative Protestant backgrounds are sometimes afraid of science or hostile toward evolution until they learn that many scientists are evangelical Christians. Writers trained in science but committed to a biblical faith can often help students sort out the religious and scientific dimensions of the public controversy over teaching evolution. Such writers are able to approach the controversy with due regard for scientific integrity but also with due respect for religious conviction:

—Charles E. Hummel's *The Galileo Connection: Resolving Conflicts between Science and the Bible* (InterVarsity Press, 1986) draws insights from the history of science for today's science/religion controversies. Hummel's small pamphlet, *Creation or Evolution?* (InterVarsity Press, 1989), does *not* insist on the choice suggested by its catchy title.

—Duane L. Thurman's *How to Think About Evolution and Other Bible-Science Controversies* (InterVarsity Press, 2nd edn., 1978), is written from a biology professor's point of view.

—John L. Wiester's *The Genesis Connection* (I.B.R.I., P.O. Box 423, Hatfield, PA 19440, 1983) presents a geologist's history of the universe and life on earth, in the outline of Genesis 1.

—Jim Brooks, a geochemist, has examined the *Origins of Life* (Lion Publishing, 1986), in a beautifully illustrated volume.

—Daniel E. Wonderly's *God's Time-Records in Ancient Sediments* (Flint, Mich.: Crystal Press, 1977) and his more recent *Neglect of Geologic Data: Sedimentary Strata Compared with Young-Earth Creationist Writings* (I.B.R.I., P.O. Box 423, Hatfield, PA 19440, 1987) both present nonradiometric evidence for an ancient earth in a winsome way to anyone inclined toward young-earth views.

—In Howard Van Till's *The Fourth Day: What the Bible and the Heavens are Telling Us about the Creation* (Wm. B. Eerdmans, 1986), a professor of physics and astronomy argues that evolution and creation are not alternatives but complementary views. Van Till and two geology colleagues, Davis A. Young and Clarence Menninga, show that there are more than two options in *Science Held Hostage: What's Wrong with Creation Science AND Evolutionism* (InterVarsity Press, 1988).

For an ongoing discussion of science/faith issues from a broad evangelical Christian perspective, *Perspectives on Science and Christian Faith* is highly recommended. Subscriptions to this quarterly journal of the American Scientific Affiliation (ASA) are $25 per year for individuals, $35 per year for schools and other institutions. From time to time, ASA has published collections of reprints from the journal, such as *Science and the Whole Person* (1985), a collection of writings about creation/evolution and many other topics by Stanford University scientist Richard H. Bube, long-time editor of the ASA *Journal*. Address of the American Scientific Affiliation is P.O. Box 668, Ipswich, MA 01938-0668. The Canadian Scientific and Christian Affiliation was incorporated in 1973 as a direct affiliate of ASA with a distinctly Canadian orientation. Its address is P.O. Box 386, Fergus, Ontario N1M 3E2.

NOTES

1. D. M. Raup, "Evolution and the Fossil Record" (letter), *Science*, Vol. 213, 17 July 1981, p. 289.
2. R. F. C. Vessot, *et al.*, "Test of Relativistic Gravitation with a Space-Borne Hydrogen Maser," *Physical Review Letters*, Vol. 45, 29 Dec 1980, pp. 2081-2084.
3. G. Gamow, "Expanding Universe and the Origin of the Elements," (letter), *Physical Review*, Vol. 70, 1946, pp. 572-573; "The Evolution of the Universe" (letter), *Nature*, Vol. 162, 30 Oct 1948, pp. 680-682.
4. R. A. Alpher and R. C. Herman, "Evolution of the Universe," *Nature*, Vol. 162, 13 Nov 1948, pp. 774-775.
5. A. H. Guth and M. Sher, "The Impossibility of a Bouncing Universe," (letter), *Nature*, Vol. 302, 7 April 1983, pp. 505-507.
6. S. A. Bludman, "Thermodynamics and the End of a Closed Universe," *Nature*, Vol. 308, 22 March 1984, pp. 319-322.
7. F. Hoyle, *Frontiers of Astronomy*, New York: The New American Library, 1955, p. 310.
8. F. Hoyle, quoted in M. W. Browne, "Scientists Expect New Clues to Origin of the Universe," New York *Times*, Sunday, 12 March 1978, pp. 1, 54.
9. H. Clemmey and N. Badham, "Oxygen in the Precambrian Atmosphere: An Evaluation of the Geological Evidence," *Geology*, Vol. 10 (3), 1982, pp. 141-146.
10. H. P. Klein, *et al.*, "The Viking Biological Investigation: Preliminary Results," *Science*, Vol. 194, 1 Oct 1976, pp. 99-105.
11. T. Dobzhansky, F. J. Ayala, G. L. Stebbins, and J. W. Valentine, *Evolution*, San Francisco: W. H. Freeman, 1977, pp. 358-360.
12. R. Shapiro, *Origins: A Skeptic's Guide to the Creation of Life on Earth*, New York: Bantam Books, 1987, p. 118.
13. W. L. Stokes, *Essentials of Earth History*, New York: Prentice-Hall, 4th edn., 1982, p. 186.
14. R. Lewin, "Alien Beings Here on Earth," *Science*, Vol. 223, 6 Jan 1984, p. 39.
15. R. Lewin, "A Lopsided Look at Evolution," *Science*, Vol. 241, 15 July 1988, p. 291.
16. D. H. Erwin, J. W. Valentine, and J. J. Sepkoski, Jr., "A Comparative Study of Diversification Events," *Evolution*, Vol. 41, 1988, p. 1183.
17. S. J. Gould, N. L. Gilinsky, and R. Z. German, "Asymmetry of Lineages and the Direction of Evolutionary Time," *Science*, Vol. 236, 12 June 1987, p. 1438.
18. R. Lewin, *Thread of Life*, Washington, D.C.: Smithsonian Books, 1982, p. 70.
19. J. W. Valentine and D. H. Erwin, "Interpreting Great Developmental Experiments: The Fossil Record," in R. A. Raff and E. C. Raff, eds., *Development as an Evolutionary Process*, New York: Alan R. Liss, Inc., 1987, pp. 97-98.
20. D. Jablonski and D. J. Bottjer, "The Ecology of Evolutionary Innovation," in M. H. Nitecki, ed., *Evolutionary Innovations*, U. of Chicago Press, in press, 1988, quoted by R. Lewin, Ref. 15, p. 291.
21. F. J. Ayala, "The Theory of Evolution: Recent Successes and Challenges," in E. McMullin, ed., *Evolution and Creation*, Notre Dame, Indiana; U. of Notre Dame Press, 1985, p. 89.
22. NOVA, "God, Darwin, and the Dinosaurs," produced for PBS TV by Larry Engle and Tom Lucas, 1989; NOVA Transcripts, P.O. Box 322, Boston, MA 02134.
23. R. Holmquist, M. M. Miyamoto, and M. Goodman, "Higher Primate Phylogeny—Why Can't We Decide?" *Molecular Biology and Evolution*, Vol. 5 (3), 1988, pp. 201-206. R. Lewin, "Conflict Over DNA Clock Results," *Science*, Vol. 241, 23 Sep 1988, pp. 1598-1600; "DNA Conflict Continues," *Science*, Vol. 241, 30 Sep 1988, pp. 1756-1759.
24. R. Lewin, *Bones of Contention: Controversies in the Search for Human Origins*. New York: Simon and Schuster, 1987, p. 319.
25. J. A. Ward and H. R. Hetzel, *Biology Today and Tomorrow*, St. Paul, Minn.: West, 1980, p. 390.
26. D. Pilbeam, "The Descent of Hominoids and Hominids," *Scientific American*, Vol. 250, (3) March 1984, p. 93.
27. S. Lipson and D. Pilbeam, "Ramapithecus and Hominoid Evolution," *Journal of Human Evolution*, Vol. 11, Sep 1982, pp. 545-548.
28. H. Nelson and R. Jurmain, *Introduction to Physical Anthropology*, St. Paul: West, 1982, p. 454.
29. R. Lewin, "Do Ape-Size Legs Mean Ape-Like Gait?" *Science*, Vol. 221, 5 Aug 1983, pp. 537-539; J. Cherfas, "Trees Have Made Man Upright," *New Scientist*, Vol. 97, 20 Jan 1983, pp. 172-178.
30. R. E. Leakey, *Human Origins*, New York: E. P. Dutton, 1982, p. 50.
31. Nelson and Jurmain, 1982, p. 337.
32. J. R. and P. H. Napier, *The Natural History of the Primates*, Cambridge, Mass.: The MIT Press, 1985, p. 29.
33. Commitee on Science and Creationism, National Academy of Sciences, *Science and Creationism: A View from the National Academy of Sciences*, Washington, D.C.: National Academy Press, 1984, p. 23.
34. Ibid., p. 24.

ILLUSTRATIONS

Front cover, 44. Drawings by Harry Blair, Greensboro, NC./6. Reprinted from *The Adventures of Huckleberry Finn* (Moby Books, Playmore, Inc., Publishers), under arrangement with I. Waldman & Son, Inc., NY. Illustration by Creig Flessel. Courtesy of Mark Twain Memorial, Hartford, CT, Elise Thall, photo archivist./ 8. Reprinted from *The Unabridged Mark Twain,* courtesy of Running Press Book Publishers, Philadelphia, PA. / 9. American Institute of Physics, Niels Bohr Library, NY./ 10., 14, 15, 16, 17. Picture Collection, California Academy of Sciences, San Francisco, CA. / 11. Reprinted from *A Short History of Astronomy* by Arthur Berry, courtesy of Dover Publications, Inc., NY./ 13. AP/Wide World Photos, NY./ 18, 19. Dept. of Library Services, American Museum of Natural History, NY: Cave paintings, Neg. No. 15038 (reindeer) and No. 317637 (bison). Kirscher photo, Neg. no. 109353./ 20, 21. Courtesy of Glen J. Kuban, North Royalton, OH, 1985, 1984./ 22, 24 (Bottom). Yerkes Observatory Photograph, University of Chicago, IL./ 24 (top). Huntington Library, San Marino, CA./ 25. Reprinted from *Red Giants and White Dwarfs,* by Robert Jastrow, Warner Books, Inc., NY, copyright © by Robert Jastrow; photo of Barrett Gallagher./ 26. Courtesy of AT&T Bell Laboratories, Short Hills, NJ./ 28, 30, 31, 41. Reprinted from *The Genesis Connection,* by John Wiester, Thomas Nelson Publishers, Nashville, TN. Used by permission./ 32 (Both). Courtesy of Walter R. Hearn, Berkeley, CA./ 33, Back cover. Courtesy of NASA, Washington, DC./ 34, 36. Drawings by Pam Rudy, D&S Composing Service./ 35 (Top and Center). Used by permission of J. W. Schopf, Center for the Study of Evolution and the Origin of Life, University of California, Los Angeles, CA; copyright © by the Precambrian Paleobiology Research Group—Proterozoic, 1986./ 37. Reproduced with permission of the Geological Survey of Greenland, Oster Volgade 10, DK-1350 Kobenhavn K., Denmark./ 38, 40 (Top). Courtesy of Anthro-Photo File, Cambridge, MA. Photo (38) by Walker./ 39. Reprinted from "The Early Relatives of Man" by E.L. Simons, *Scientific American* (July 1964). Copyright © 1964 by Scientific American, Inc. All rights reserved./ 40 (Bottom), 42. Courtesy of Institute of Human Origins, Berkeley, CA./ 43 (Both). Courtesy of Helen Martin./ 48. Courtesy of the Sierra Club, San Francisco, CA.

Back cover photos. Top: Spiral galaxy NGC 6946 (L); "Veil" nebula, NGC 6992-5 (center); Sagitarius star cloud (R).
Bottom: Echo satellite trail in Milky Way.

Book design by Virginia Hearn.
Typography by D & S Composing Service.
Printed by Science Press.

Addendum: Classroom Exercises

Science teachers using earlier versions of this guide asked for practical exercises to teach **critical thinking skills** related to its subject matter — lessons that go beyond passive acceptance of information.

In 1990 an outstanding exhibit called "Life Through Time: The Evidence for Evolution" opened at the California Academy of Sciences in Golden Gate Park in San Francisco. One display in that exhibit, "The Hard Facts Wall," has provided material for an exercise on fossils and on the inferences that can be drawn from fossil sequences.

We encourage teachers to make copies of the illustrations provided, and to design their own lesson plans using them. The three suggested lessons have been taught most successfully in one 90-minute class period. For a single 50-minute period, in the first half combine Lessons 1 and 2 by having students do their classifying and plotting at the same time.

Student Learning Objectives

1. To identify fossils (represented by drawings on cards) by their external features and then to arrange them into appropriate classification categories (e.g., into phyla).

2. To plot data in a systematic fashion by placing the fossil cards on a graph of the ages of strata in which the fossils were found (given on each card).

3. To derive appropriate inferences from the plotted data.

4. To be able to distinguish between evidence and inference regarding a given set of data, thus developing skill in critical thinking.

Fossils, the "Hard Facts"

Fossils are remnants of past life, such as dinosaur bones, ancient clam shells, footprints of a long-extinct animal, or impressions of a leaf in rock. Real fossils (in one or two cases, casts of fossils) belonging to six major taxonomic categories constitute the *facts* of The Hard Facts Wall. For this exercise, in each category the oldest and the most recent fossils, plus three others, have been selected.

This exercise is best appreciated by students who have *handled* or at least *closely observed* some actual fossil specimens. If at all possible, take your class to visit a museum where fossils are displayed. Teachers in California are encouraged to plan a visit to the impressive Life Through Time exhibit, including The Hard Facts Wall. (Entrance to the California Academy of Sciences is free on the first Wednesday of every month.)

The following lesson plans were developed by Kevin Callaway, a biology teacher at Hueneme High School in Oxnard, California. After initial classroom testing, Kevin presented the results in a paper in the Education Section of the June 1992 Annual Meeting of the Pacific Division of the AAAS (American Association for the Advancement of Science). ❏

Lesson Plan No. 1:
Classification of Fossils by External Features

Teacher Preparation

1. Make enough copies of p. 53 on good paper or light card stock for each group of 2-3 students. Cut each sheet into a set of cards, shuffle them, and place in an envelope for each group.

2. Make a copy of p. 56 for each group, for students to check their classification.

3. Be prepared to discuss scientists' reasons for classifying organisms into categories based on similar characteristics. (VERTEBRATES are a subphylum of CHORDATA; the other five categories in the museum display are invertebrates; CORALS are a class of the phylum CNIDARIA.) Write the names of the six categories on the board and outline the distinguishing characteristics of each group. An alternative to using the board is to make a transparency of p. 53 and use an overhead projector, at first covering all but the category names.

PROBLEM

To place fossil cards into appropriate taxonomic groups on the basis of similar external characteristics.

MATERIALS

1. Thirty fossil cards.

2. Museum presentation graph: Life Through Time: The Evidence for Evolution (p. 56).

PROCEDURE

1. Place students into small groups (2-3) and provide each group with a set of fossil cards.

2. After discussing the characteristics of each of the six taxonomic categories, have students sort the cards into the proper categories, five cards in each.

RESULTS

Some cards will be easy to classify, but students will have trouble with others because of the limitations of sketches to convey actual characteristics. Several of the oldest fossils, such as the brachiopod *Acrotreta,* are essentially unclassifiable by observation. After the students have done their best, pass out copies of p. 56 for them to check their own classification against the museum's. On their own, many students will place the fossils in each group in chronological order; or the teacher can suggest this, anticipating Lesson 2.

CONCLUSION

Animal life in the ancient past can be grouped by external features in the same way as present living things. Some ancient forms of animal life closely resembled modern animals, others less so. ❑

Lesson Plan No. 2:
Plotting Fossil Cards by Age on a Graph

Teacher Preparation

1. Make a copy of pp. 54 and 55 for each group from Lesson 1 and tape the two pages together to form an approximately 10" x 14" graph that will lie flat.

2. Note that on the y-axis "time" is divided into 100-million-year steps from 600 mya (million years ago) at the bottom to the present at the top. You may want to discuss how geological strata have been assigned *relative* dates (e.g., from the principle that younger layers generally overlay older ones) and *absolute* dates (from radioactive decay measurements).

3. Prepare a transparency of p. 57 for overhead projection (or make a copy of p. 57 for each group).

PROBLEM

To place the fossil cards at the proper age on a graph, accurately plotting empirical data for pattern analysis and interpretation.

MATERIALS

1. Materials from Lesson 1: fossil cards in proper taxonomic categories plus a copy of p. 56 (museum presentation).

2. Graph, from taping together copies of pp. 54 and 55.

3. Copy of p. 57 (empirical plot), unless prepared as a transparency.

PROCEDURE

1. Have students turn museum presentation (p. 56) face down, Students tempted to "peek" at p. 56 should be cautioned that, in science, copying can be counter-productive.

2. Have students fold their fossil cards like little pup tents, with the age readable from the front. Then have them "plot" the fossil data on the large graph by arranging the cards along the appropriate vertical line (under CORALS, MOLLUSKS, ARTHROPODS, BRACHIOPODS, ECHINODERMS, and VERTEBRATES) at the matching age on the graph.

3. When they have finished, have them turn over the museum presentation and compare it to their own arrangement of the data. Finally, refer students to the empirical plot (p. 57).

RESULTS

Student plots should look like p. 57. In a small class the teacher can visit each group, handing each a copy of p. 57 and checking their work against it; in a larger class p. 57 can be projected for students to check their own work.

CONCLUSION

Empirical plots of the fossil data by students show a pattern different from that of the museum display. Students can see that in the museum display, some younger fossils are arranged so that they appear below older ones. Since the two different graphs could lead to different inferences, congratulate your students for being "good empirical scientists." ❑

Lesson Plan No. 3:
Exploring Evolutionary Relationships

Teacher Preparation

1. Make transparencies for overhead projection of diagrams on pp. 58, 59, 62, and 63.

2. Read "The Critical Importance of Critical Thinking" on pp. 60-61. Compare the photograph of The Hard Facts Wall (p. 61) with the diagrams on pp. 58-59 and 62-63.

3. Be prepared to define and discuss the terms *evidence* (e.g., "something that tends to prove; ground for belief") and *inference* (e.g., "conclusion derived by reasoning").

PROBLEM

To arrive at appropriate inferences from evidence in the form of a fossil data-set graphed by students, and also from a museum presentation of the same data.

MATERIALS

1. Student-constructed graph from Lesson 2.

2. Museum presentation, Life Through Time: The Evidence for Evolution (p. 56); Life Through Time (p. 57) unless the latter is used as an overhead projection instead.

PROCEDURE

1. Encourage students to recognize that they have plotted the data correctly, despite any differences with the museum presentation. For many students, this accomplishment builds self-confidence and self-esteem.

2. Show overhead projections of pp. 58 and 59, asking students to point out examples of evidence and inference. Use projections of pp. 62 and 63 to illustrate how previously held assumptions can influence the presentation of "hard facts."

RESULTS

Students should be able to see that the two conflicting graphs can lead to different interpretations. The conflicts will be most obvious to students who were strongly influenced by first seeing the museum's way of graphing the data.

CONCLUSION

Data may be presented in different ways because of previously held assumptions. Valid scientific inferences depend on proper presentation of data. Inferences based on assumptions rather than on evidence may mislead us. In the museum display, magnifying glasses are placed at "branch points," but there are no fossils under them. What are the implications of the absence of those fossils?

DISCUSSION QUESTIONS

(Preferably, have students write their answers in essay form. If assigned as homework, provide students with take-home copies of pp. 56 and 57.)

1. Describe how the museum presentation differs from your graph.

2. Did inference affect your presentation of the data? Explain.

3. Did inference affect the museum's presentation of the data? Explain.

4. Why do you think the museum presented the data differently? ❏

Corals	Mollusks	Arthropods	Brachiopods	Echinoderms	Vertebrates
Pocillopora damicornis (recent specimen)	Scallop (*Swiftopecten swiftii*) 1 mya	Barnacle (*Balanus*) 12 mya	*Laqueus vancouverensis* (recent specimen)	Sand dollar (*Dendraster*) 3 mya	Stickleback (*Gasterosteus dorrysus*) 5 mya
Siderastrea 20 mya	Snail (*Turritella uvasana*) 45 mya	Lobster (*Palinurina longipes*) 152 mya	*Cretirhynchia plicatilis* 85 mya	Seastar (asteroid) 80 mya	Fish (*Diplomystus*) 50 mya
Heritschioides 270 mya	Clam (*Trigonia costata*) 170 mya	Trilobite (*Calymene*) 400 mya	*Neospirifer* 270 mya	Brittle star (*Ophiopinna elegans*) 165 mya	Coelacanth (*Holophagus*) 152 mya (cast of fossil)
Zaphrentis 385 mya	Snail (*Platyostoma*) 425 mya	Trilobite (*Ogygopsis klotzi*) 530 mya	*Paraspirifer bownockeri* 380 mya	Crinoid (*Aorocrinus immaturus*) 355 mya	Coelacanth (*Diplurus newarki*) 220 mya (cast of fossil)
Favistina stellata 440 mya	Monoplacophoran mollusk (*Scenella*) 530 mya	Trilobite (*Olenellus thompsoni*) 570 mya	*Acrotreta* 530 mya	Echinoderm (*Helicoplacus gilberti*) 550 mya	Heterostracan vertebrate (*Astraspis*) 460 mya (fossil bony plates)

Life Through Time

	Corals	Mollusks	Arthropods

Millions of Years Ago (MYA)

Present

100

200

300

400

500

600

Life Through Time

Brachiopods	Echinoderms	Vertebrates	Millions of Years Ago (MYA)
			Present
			100
			200
			300
			400
			500
			600

Life Through Time: The Evidence For Evolution

Museum Presentation

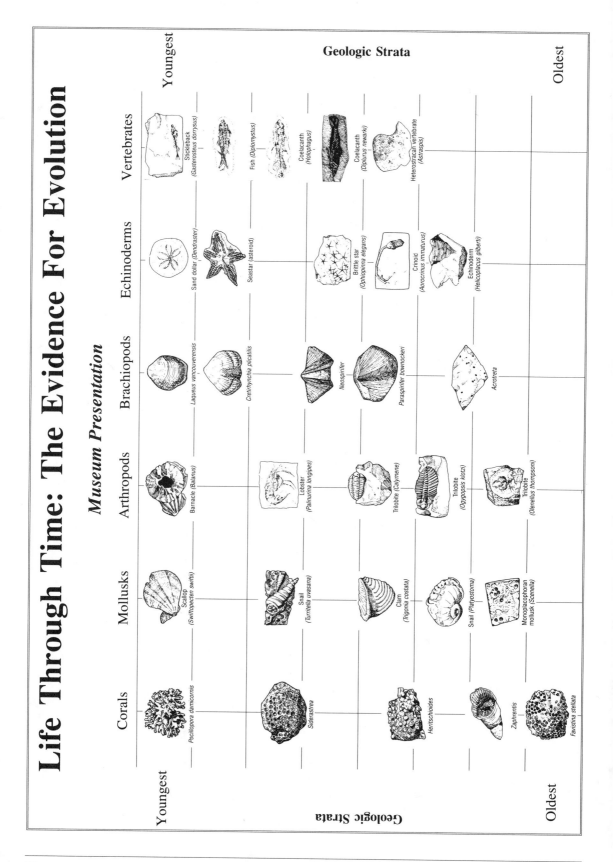

Geologic Strata

Youngest

Oldest

Geologic Strata

Youngest

Oldest

Corals

Mollusks

Arthropods

Brachiopods

Echinoderms

Vertebrates

Pocillopora damicornis

Siderastrea

Heritschioides

Zaphrentis

Favistina stellata

Scallop (*Swiftopecten swiftii*)

Snail (*Turritella uvasana*)

Clam (*Trigonia costata*)

Snail (*Platyostoma*)

Monoplacophoran mollusk (*Scenella*)

Barnacle (*Balanus*)

Lobster (*Palinurina longipes*)

Trilobite (*Calymene*)

Trilobite (*Ogygopsis klotzi*)

Trilobite (*Olenellus thompsoni*)

Laqueus vancouverensis

Cretirhynchia plicatilis

Neospirifer

Paraspirifer bownockeri

Acrotreta

Sand dollar (*Dendraster*)

Seastar (asteroid)

Brittle star (*Ophiopinna elegans*)

Crinoid (*Aorocrinus immaturus*)

Echinoderm (*Helicoplacus gilberti*)

Stickleback (*Gasterosteus doryssus*)

Fish (*Diplomystus*)

Coelacanth (*Holophagus*)

Coelacanth (*Diplurus newarki*)

Heterostracan vertebrate (*Astraspis*)

Life Through Time

Empirical Plot

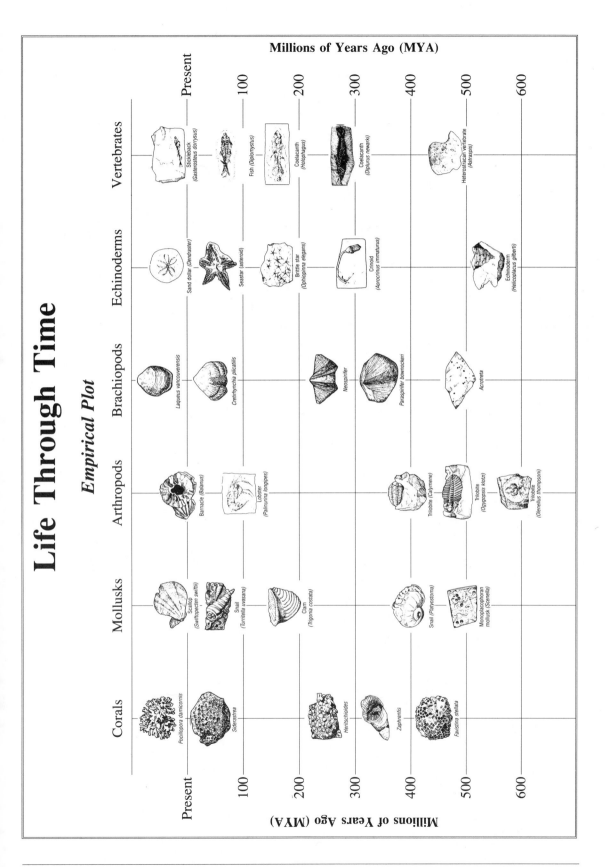

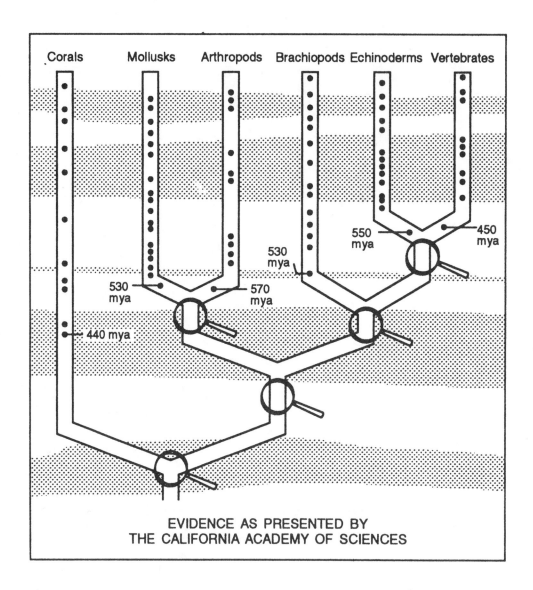

Corals　Mollusks　Arthropods　Brachiopods Echinoderms Vertebrates

550
mya

450
mya

530
mya

530
mya

570
mya

440 mya

EVIDENCE AS PRESENTED BY
THE CALIFORNIA ACADEMY OF SCIENCES

Diagramatic representation of The Hard Facts Wall. In the museum, the dates for the fossil specimens are displayed on an adjoining wall. In this diagram, dates for the oldest specimens are shown to highlight the inaccurate placement of fossils. Compare this diagram to the one on p. 59. (This diagram and the one on p. 59 are reprinted with permission from M. Hartwig and P. A. Nelson, *Invitation to Conflict: A Retrospective Look at the California Science Framework,* Access Research Network, Colorado Springs, 1992.)

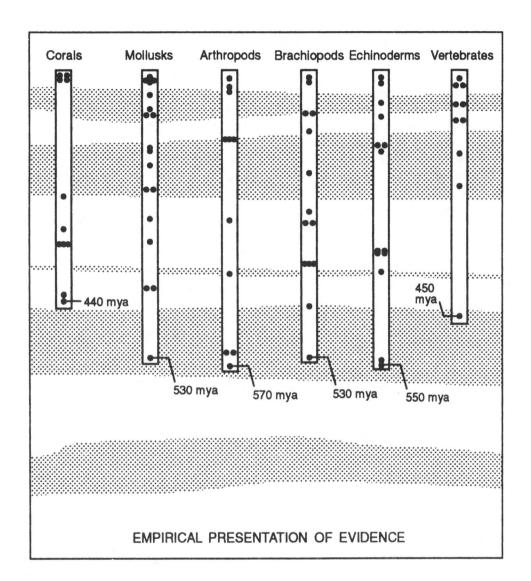

Corals Mollusks Arthropods Brachiopods Echinoderms Vertebrates

440 mya

450 mya

530 mya   570 mya   530 mya   550 mya

EMPIRICAL PRESENTATION OF EVIDENCE

Diagramatic representation of a more empirical arrangement of fossil evidence from The Hard Facts Wall. Note that the oldest mollusk, arthropod, brachiopod, echinoderm, and vertebrate fossils actually belong *below* the oldest coral fossil, not above it, as in the museum display diagramed on p. 58. Note also the "clumping" of the fossils in the empirical plot above; some vertical spreading out of the fossils by the museum was probably done simply for better display of each specimen.

The Critical Importance of Critical Thinking

To think critically does not mean to cultivate a negative attitude, but to ask thoughtful questions. Adolescents grow by learning to question authority — parental, religious, scientific — without automatically rejecting it.

In order to "think like a scientist" a person needs to (1) distinguish between evidence and inference, and (2) appreciate the difference between facts, hypotheses, and theories. Even students who will never become scientists can grasp these distinctions.

Most museum visitors probably accept The Hard Facts Wall as an authoritative presentation of scientific evidence. On the other hand, young people beginning to think like scientists ought to ask: Is everything here really a hard fact?

Thinking Critically about Evolution

Ideological and political battles are still being fought over teaching evolution in public schools. You can keep such polemics out of your classroom and still teach your students to think critically about evolution. A 1991 resolution from the American Scientific Affiliation* urges science teachers to be clear about what they are teaching — by carefully defining the terms *evolution* and *theory of evolution* and by using those terms in a consistently scientific manner.

Relying on "The Meanings of Evolution" outlined by biologist Keith Stewart Thomson (*American Scientist,* Vol. 7, pp. 529-31, Sept-Oct 1982), the ASA resolution is directed not only to teachers but to everyone directly concerned with public education in science.

Evolution should be taught as fact when it is defined simply as "change through time." Obviously, this broadest of Thomson's three definitions is not limited to biology. People speak of the evolution of our solar system, evolution of the computer, evolution of the short story, and so on.

The fossil evidence does show a changing biological pattern, but observant students may also see on The Hard Facts Wall evidence of things that have *not* changed. The major categories have not changed, for example. Since the beginning of the Cambrian over 500 million years ago, many animal phyla have disappeared — yet new ones have not appeared. Why? Within the phyla, the trilobites are gone, new animals have appeared, and some forms do seem to have been modified—yet others look essentially the same after hundreds of millions of years. If some things change, but some do not, is it accurate to say that everything evolves?

Evolution should be taught as hypothesis when it is taken to mean that "all organisms are related through common ancestry." Although such a relationship is plausible, it is nonetheless still "hypothetical." Classifying life forms by common characteristics is relatively easy. Establishing genetic links "across the board" is far more difficult. Common genetic ancestry of the animal phyla is not an established fact but an *inference* based on a certain amount of *evidence*.

Evolution should be taught as theory when the term refers to the way (or ways) Darwin and his successors have sought to explain a (factual) historical pattern and a (hypothetical) historical process. Mutation and natural selection have been demonstrated to produce a limited amount of biological variation. But whether they (or other proposed mechanisms) can adequately account for the whole history of living forms is still "theoretical."

Although "hypothetical" and "theoretical" are sometimes used interchangeably even in scientific writing, scientists generally make a distinction between hypothesis and theory. Scientists are continually trying to confirm or reject experimentally their tentative working hypotheses. Like such hypotheses, scientific theories are imaginative proposals. But because they are constructed to account for a wider range of facts and hypotheses, theories are broader in scope than hypotheses. In science, theories and hypotheses are not just "anybody's guess"; they are inferences, some better supported by evidence than others. Teaching science without hypotheses and theories is hardly teaching science at all.

---

* The background and full text of the ASA resolution, "A Voice for Evolution *As Science*," are printed on the blue card in the front of this book.

Teaching Evolution *As Science*

These exercises will not introduce your class to theories of evolution, but they will help students face some of the facts of evolution and ponder the major evolutionary hypothesis: common genetic ancestry.

Besides emphasizing clear definitions, the ASA resolution also recommends three ways to make science classes more stimulating. The three suggestions, with applications to these exercises on critical thinking, are as follows:

Present well-established data and conclusions forcefully. The fossils on The Hard Facts Wall clearly demonstrate that biological change has occurred over geological time. Life on this earth has a long history.

Distinguish clearly between evidence and inference. On The Hard Facts Wall, the fossils and their ages constitute evidence of life's history. The lines connecting the taxa, implying common ancestry, are examples of inference. To reinforce the distinction for your students, point to the absence of any fossils under the magnifying glasses placed by the museum at points where hypothetical common ancestors would be expected. Using transparencies of pp. 62 and 63, you can illustrate the principle that basing conclusions on inference rather than on evidence is a risky and unscientific procedure.

Discuss unsolved problems and open questions candidly. The origin of the animal phyla is an open scientific question, discussed on pp. 34-37 of *Teaching Science*. With these exercises, you can show your students how the ideological framework of The Hard Facts Wall sets forth a classic prediction of Darwinian theory. That is, the neo-Darwinian mechanism of natural selection operating on random mutation should give rise to major new body plans (i.e., to new phyla) as one moves upward through the geological strata.

As the exercises show, however, in designing the display the museum chose to misrepresent the placement of fossil data to make them better fit that hypothetical framework. One of the "hard facts" of paleontology is that virtually all of the animal phyla appear in the fossil record in one brief geological period — in contradiction to the Darwinian prediction of new phyla appearing over geological ages.

What will your class make of this unsolved scientific problem? Perhaps in your class are some future scientists who will be inspired to search for new mechanisms to explain the non-Darwinian pattern of the fossil record. ❏

The Hard Facts Wall, a display in the exhibit Life Through Time: The Evidence for Evolution, California Academy of Sciences, Golden Gate Park, San Francisco. This particular display does not include a time scale; relative time is represented by horizontal sedimentary layers forming a backdrop against which the fossils are mounted. Individual fossils in each vertical column are identified by numbers keyed to a wall chart at left, giving the species name, the general location where the specimen was found, and its geological age. In some other displays in the Life Through Time exhibit, magnifying glasses are mounted over tiny fossils; on The Hard Facts Wall, the magnifying glasses have no fossils beneath them. (The bent shape of the backdrop, neutral tones of the fossils and background, and glass pane in front of the display make it difficult to get a good photograph. This attempt was made in 1991.)

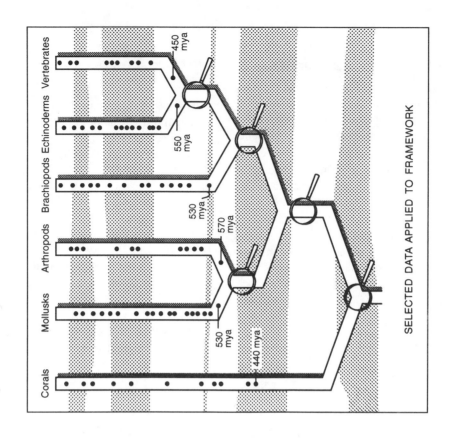

SELECTED DATA APPLIED TO FRAMEWORK

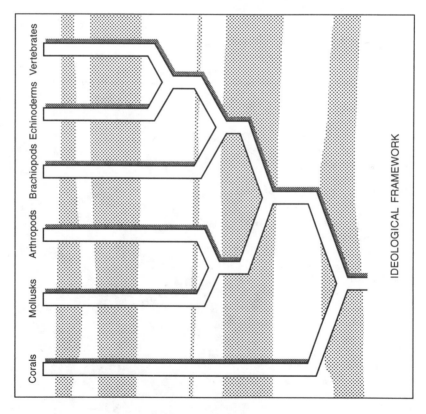

IDEOLOGICAL FRAMEWORK

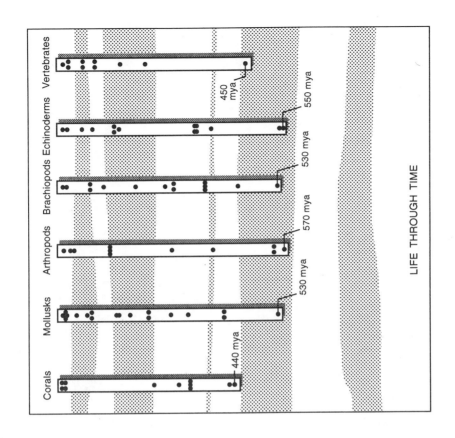

LIFE THROUGH TIME

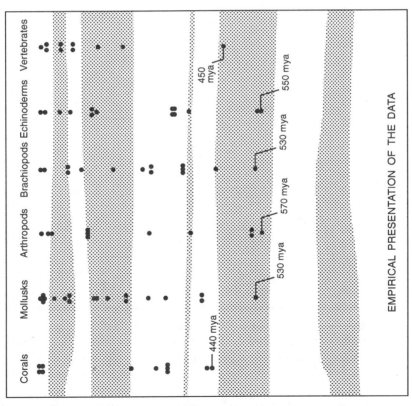

EMPIRICAL PRESENTATION OF THE DATA

Notes

American Scientific Affiliation Executive Council, 1993; Elizabeth M. Zipf, President; Fred S. Hickernell, Vice President; Raymond H. Brand, Secretary-Treasurer; Kenneth J. Dormer, Past President; and David L. Wilcox. Executive Director: Robert L. Herrmann. ASA Committee for Integrity in Science Education, 1993: John L. Wiester, Chair; David Price; and Walter R. Hearn. Design by Patricia Ames. Artistic consultant: Jay Flom. Editorial and design consultant: Virginia Hearn.